Ultramini Morphable T-Ray FR1 Zone Metatransceiver

Clara Belle

Ultramini Morphable T-Ray FR1 Zone Metatransceiver

First Edition November 2025

Written by Clara Belle

Table of Contents

TABLE OF CONTENTS

Chapter I

CHAPTER I

INTRODUCTION

1.1 WIRELESS COMMUNICATIONS

With the progress of new technologies, the expansion of wireless communications has brought about a new transition in the current communication environment. It is evolving day by day in response to the demands for new advanced communication technologies. People's real activities are becoming more dependent on high information and data speeds. The entire communication history can be divided into three sections: the pioneer age, the pre-cellular age and the cellular age. Maxwell conducted experiments to construct equations for electromagnetic (EM) wave propagation during the pioneer period, which lasted from 1860 to 1921. Around 1865, these fundamental laws of EM waves have been developed. According to Maxwell, electric and magnetic fields are created by charges, currents, and field changes. These equations show how fluctuating the electric and magnetic fields, known as electromagnetic waves, which moves at the speed of light.

Many renowned scientists demonstrated various radio wave experiments during this time period. Heinrich Hertz carried out an experiment. According to this, the first experiment involving the transmission and receiving of electromagnetic waves have been accomplished. The EM waves propagating from oscillator sparks will generate or produce a current in the loop. Around 1893, Nikola Tesla postulated that energy may be transmitted from one location to another via electromagnetic waves. Following to this theory, another scientist Alexander Popov, who conducted tests about 1895 and released a report on wireless lightning detectors constructed with the coherer.

In Calcutta at the time, physicist Jagadish Chandra Bose demonstrated radio waves. Milli-meter waves and the concept of semiconductor junctions were the two most important achievements at the time. Guglielmo Marconi pioneered radio telegraphy and demonstrated that radio transmission could be used to deliver messages over 2200 miles across the Atlantic. Claude Shannon then proposed a Mathematical Theory of Communication in 1948 and is widely regarded as the father of information theory, which revolutionised the communication systems.

In the pre-cellular era, mobile communication was originally implemented via electromagnetic waves (1921 to 1980). Card communicators were widely used during this time period. These, however, are not linked to the public voice telephone network. These networks were mostly self-contained. There were issues with handoff, and the zone of coverage in this region was limited. During World War II, there was a major stimulant to radio communications around 1940, which led to a massive growth in the development of hand-held devices using semiconductor transistors.

After that, mobile communication was established and widely implemented utilising advanced technology, and this era is known as the third or cellular era. During this time, analog cellular communication technologies are being deployed. The analog communication system was preceded by digital cellular communication systems and the introduction of public land mobile telecommunication systems, before the IMT-2000 was developed in 2010 [1]. After 2010, there were deployments of 4G networks, and active work is being done on a 5G system [2].

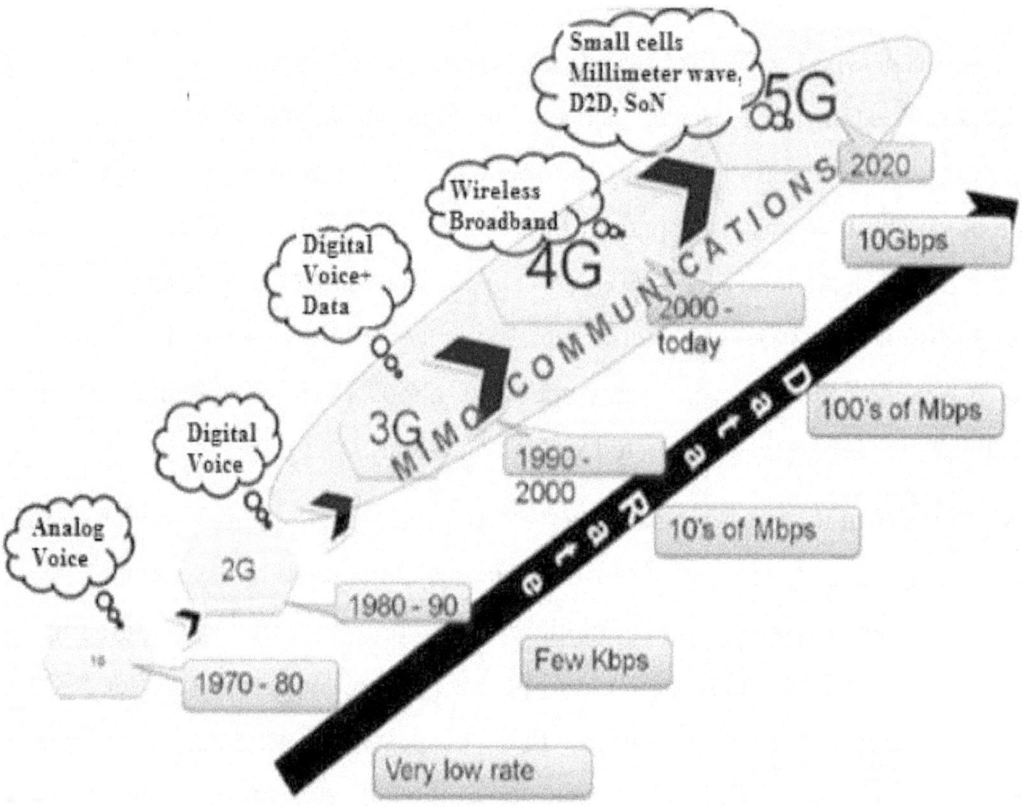

Figure 1.1. Evolution of Wireless Communication Systems

The Progress of Wireless Communication Systems is given in Figure 1.1. First Generation 1G (1970-1980s) is fundamentally for analog communication, with speech being the primary deciding element in system design and no data transfer provided [3]. Second Generation 2G (1980-1990) systems are distinguished by digital voice and thus support digital communications. However, voice will continue to be the most significant traffic handled. During the transition from 2G to 3G, a period known as 2.5G occurred during which data was introduced to be transmitted via mobile network systems and would carry data within the channel used to carry voice signals. Third Generation 3G, a digital technology, was created between the years of 1990 and 2000, and its implementation began in the early years of 2001. It established distinct voice and data routes [4-5]. With the widespread usage of the internet, the demand for data services is fast increasing, necessitating a high data rate in the system. An increasing need for data-centric services for all purposes, from entertainment to scientific and administrative purposes, which resulted in the development of a fourth generation 4G mobile communication infrastructure that is basically designed to transmit data traffic. Voice over IP, sometimes known as VOIP, is commonly used in 4G networks [6].

Fifth Generation 5G is being developed as the demand for faster data rates grows. New generation wireless communication devices enable ordinary people and machines to communicate in a completely mobile manner. Previously, the data rate required was extremely low. To enhance the data rate, newer technologies with new applications are integrated into the system, and thus the technology has progressed beyond 4G into the 5G era. Among the numerous things moving forward with 4G are the creation of tiny cells, device-to-device communication, self-organizing networks, and so on. Millimeter wave communication and massive MIMO, among other technologies, are likely to be major features of 5G. All 5G systems are expected to utilise MIMO communication systems, which is one of the key technological differences from previous systems [7-9].

This chapter is organized into various sections as follows: Section 1.1 starts with the introduction of wireless communication systems. Section 1.2 describes about the reconfigurable antennas. Section 1.3 presents advanced MIMO system introduction followed by Section 1.4 applications of MIMO systems. Section 1.5 offers antenna diversity techniques followed by Section 1.6 having mutual coupling and MIMO antenna performance parameters. Section 1.7 discusses about the simulation tools followed by Section 1.8 on antenna fabrication and experimental analysis. Section 1.9 presents the objectives and scope of the work. Section 1.10 describes the organization of thesis.

1.2 RECONFIGURABLE ANTENNAS

There are several areas and applications which require antenna to transmit and receive signals. Antenna is an important element of many communication systems. Generally wireless communication systems are used for different applications. Some of the characteristics such as radiation pattern, polarization and frequency are considered in simple antenna systems. There are a lot of emerging applications in modern communication systems which requires an advanced and intelligent antenna systems. So, antennas with additional functionalities are required for industries. A good option for this type of antenna systems is described as Reconfigurable antennas, as they can be reconfigured to operate over different frequencies or pattern or polarization. It can offer multiple functions or applications into a single device. There are different technologies that can be used to convert antennas into reconfigurable antenna. The antenna can be reconfigured by modifying its structure and the switching states that are used. In designing a reconfigurable antenna, there are a lot of challenges. Achieving reconfiguration in a required manner without affecting any other parameters is the toughest issue.

1.2.1 Basic Approaches to Generate Different Reconfigurable Antennas

There are diverse types of reconfigurable techniques. To accomplish the reconfigurability in antennas, some basic principles are needed to achieve.

- The surface current distribution on the antenna or on its structure needs to be changed in order to have frequency-reconfigurable antennas.
- To have a pattern reconfigurable antenna, the slots, radiating edges, current distribution, or feeding network must be modified.
- To have polarisation reconfigurable antennas, the surface structure, current distribution, or feeding network of the antenna must be altered.
- A particular antenna may have several reconfigurable capabilities as a result of the combination of many reconfigurability principles.

- It is essential to keep in mind that altering the current distribution will alter how electromagnetic fields are distributed throughout the antenna's aperture.
- Other antenna parameters may be impacted and changed if one of the antenna properties is altered.

1.2.2 Merits of Reconfigurable Antennas

1. One antenna can support multiple wireless standards. The following are the advantages gained by having this benefit:
    - Radio technology costs will be reduced as a consequence of the intricacy of the technology and the incompatibilities of communication systems.
    - The volume used will decrease.
    - Integration will be informal.
    - There will be effective isolation between various wireless protocols.
2. The amount of front-end processing required will be reduced. The following advantages have been gained as a result of this point:
    - There will be no need for front-end filtering.
    - Out-of-band rejection will be excellent.
3. Attributes for multitasking:

- It can function as an array or a single element.
- It can operate in either narrowband or wideband mode.
- If the mission is modified, the functionality can be adjusted.

4. Enhances the sustainability of IoT applications and the effectiveness of Massive MIMO systems.

1.2.3 Different Reconfiguration Techniques

1.2.3.1 Electrical

Most of the time, this technique is used. Switches can be used to connect a slot and to separate components which achieve specialised radiation properties such as circular polarisation, or encompass the extent of an antenna element. Slots are now connected to alter surface current distributions. This technology is appealing because it is simple to incorporate switches inside the antenna. Typically, these switches are employed to reroute the surface current. Some of the switches can perform using this technique are used by RF-MEMS, PIN diodes, Varactors etc. The electrical process uses both discrete and continual tuning techniques. The application is simple and focuses on several factors in the field of study. DC source and biassing circuits are required for these antenna components to function. As a result, this antenna is dependent on a DC electrical supply and electronic switching components that degrade the antenna's process and performance.

1.2.3.2 Optical

Using photoconductive elements, this technique is possible. When a photoconductive material is irradiated by light of the suitable wavelength, electrons shift from the valance to the conduction band, causing the material to conduct current and the switch to shift from the OFF to ON state. A laser might be built inside the antenna, or light could be produced from an optical cable in the design. In optical reconfiguration, photoelectric switches are utilised instead of electric switches because they give less disturbance, higher separation, and faster switching speeds. This enables one to adjust the antenna operational bandwidths and radiation patterns optically.

1.2.3.3 Mechanical

This technique involves modifying the design of the antenna, which causes current to be redistributed and radiation characteristics to vary. This approach has the advantage of not requiring biassing lines, lasers, or switching. However, they have a very poor response time, and are expensive, large in size, use a lot of power, and are challenging to integrate with an antenna. The use of this technology will also be constrained by device limitations, which might make it impossible to adjust the antenna mechanically.

1.2.3.4 Switching using Tunable Materials

It is possible to accomplish by using ferrites or liquid crystals. The dielectric constant of a liquid crystal changes as the applied voltage changes. The permittivity or permeability of ferrite is altered by the applied static electric or magnetic field. The fundamental benefit of this technology is that smart materials are small in size, but the drawback is that they have low efficiencies, especially for ferrites.

1.2.4 Types of Reconfigurable Antennas

Reconfigurable antennas can modify their features, which include frequency, polarization, pattern and bandwidth. Changing the current distribution or the design of the antenna structure changes the field close to the antenna structure as well as the radiated distant field. Reconfigurable antennas are classified into four categories based on the parameters that can be changed. Multiple antennas can have two or more attributes separately modified. Compound antenna reconfiguration is a significantly more developed subject with greater work put into merging many functions in a single reconfigurable antenna. Many new concepts are emerging in antenna applications, particularly for current communication devices, owing to the concept of multiple reconfigurability. Although singular reconfigurability is used, multiple- reconfigurability is far more complex and advanced. The interdependence of antenna parameters makes it difficult to adjust numerous parameters at the same time. Several scientific attempts, however, have been admitted in building compound reconfigurable antennas. In general, both multiple and single reconfigurability approaches are used where careful consideration of the design and control mechanisms is required. There have been numerous notable efforts on multi-reconfigurability antennas that have been examined for frequency and pattern reconfigurability, frequency agility, polarisation diversity, and pattern and polarisation reconfigurability.

1.2.5 Applications of Reconfigurable Antennas

Wireless communication has advanced to a very high application in the new generation that can respond to the environment based on the surrounding conditions. Cognitive radio systems, satellite communications, military applications, MIMO systems, and medical applications are some of the examples used for both terrestrial and astronomical applications. Some of them are presented as follows:

1.2.5.1 Frequency Reconfigurable Antenna for Cognitive Radio System

As a result of clear rapid expansion of the communication system, the demand for frequency bands is rising, resulting in RF spectrum scarcity. The main reason of RF spectrum scarcity is the inadequate spectrum distribution. As more frequency bands are utilised, this becomes more prevalent. Cognitive radio, which increase spectrum efficiency, are employed to address future communication issues. By sensing and analysing available channels before beginning transmission, this can utilise unoccupied space across a wide frequency range. The evolution of this system poses a significant challenge to antenna development. In general, the advent of CR and spectrum allocation present new problems to frequency reconfigurable antennas since the antenna can adjust the frequency without losing gain or radiation qualities.

One approach is to add a reconfigurable filter within the antenna design, known as filtenna, which can be implemented at the feeding line or on its ground plane. In order to operate in several bands, the majority of the reconfiguration mechanisms are built within the ultra-wide band (UWB) antenna. It is possible to accomplish this by employing switches such as ideal and optical switches, PIN diodes, varactor diodes, linear actuators, and stepper motors. A cognitive radio can observe the channel and identify its idle frequencies. Idle frequency, in essence, concerns with white space or unoccupied space. In reconfigurable antennas, the idle frequency is quite significant. After identifying the vacant space, the reconfigurable antenna can adapt its operation to meet the requirements and therefore boost efficiency. The cognitive radio has the benefits of allowing the device to decide and configure. It may self-realize the mode of communication chosen.

1.2.5.2 Pattern Reconfigurable Antenna for MIMO Antenna Systems

MIMO uses many antennas at both the transmitter and receiver front ends to continually broadcast different information, improving communication spectral efficiency in a multipath environment. In response to changing channel conditions and requirement analysis, a MIMO system may adjust the modulation level, coding rate, and transmission signalling systems. Due to the obvious substantial propagation loss of Millimeter waves in 5G systems, MIMO and beamforming techniques are crucial for enhancing spectral efficiency while providing cost-effective and dependable coverage. Beamforming is one of the uses of numerous radiating devices that transmit the same signal at the same phase and wavelength, combining to form a single antenna with a more concentrated stream formed by amplifying waves in a defined direction.

Beam steering is a technique that involves adjusting the phase of the input signal on all radiating elements. The signal can be focused towards a specified receiver via phase-shifting, and the antenna can use radiating elements with a shared frequency to drive a single beam in a defined direction. MIMO systems benefit from the usage of reconfigurable antennas. It employs multiple antennas at both the transmitter and receiver front ends to simultaneously broadcast several data streams. In MIMO systems, reconfigurable antennas are utilised to improve channel reliability, capacity, and system performance.

1.2.5.3 Reconfigurable Antenna for Satellite Applications

As a result of demand for wireless and space applications, reconfigurable antennas for satellite applications have been developed. In this technology, the antenna radiation pattern is reconfigured to help coverage zones maintain high data rates at multiple frequency bands. The future data network in metropolitan regions is not well developed, but it can be circumvented via a satellite communication system. Google launched the Loon project, which includes the deployment of flying balloons to give internet access everywhere. The flying balloon is less expensive, but its lifespan is limited, whereas the satellite is more expensive and has a much longer lifespan. A steerable antenna layout is required for satellite antennas. It is used to align the antenna of a low-Earth-orbit satellite with a geostationary satellite. On a satellite platform, numerous applications for beam-steering and antenna shaping are presented. In the case of satellites and balloons, both nodes should be linked to establish a backbone network that would act as a networking platform for users. It depicts communication between low-Earth orbit and the station, which is either in geostationary orbit or on the ground. Beam steering is required at the antenna nodes.

1.2.5.4 Reconfigurable Antenna for Biomedical Applications

Due to the advancements in microelectronic devices, researcher's interest in communication, biotelemetry, and other fields has expanded. Various biomedical gadgets have been developed to interface with the human body in order to monitor and diagnose disorders. Implantable medical devices have been implemented in biotelemetry communication to communicate wirelessly between implanted devices and external units. Implantable medical devices are applied in industrial, scientific, and medical bands. Biomedical technology is now commonly regarded as a technique of enhancing medical diagnosis and treatment, as well as an academic discipline. Glucose monitoring, insulin pumps, deep brain simulations, and endoscopy are just a few of the medical applications that can benefit from a distant monitoring system and a body implanted unit today. Body implantable devices are being studied extensively for human applications. Antennas can be placed in humans to build a bio-communication system between medical equipment and external instruments for short-range biotelemetry applications. Aside from the obvious benefits to the healthcare system afforded by body implanted devices, economic considerations are also important. Remote monitoring technologies aid in disease diagnosis and benefit the hospital at home by lowering hospitalisation time.

1.3 MULTIPLE-INPUT MULTIPLE-OUTPUT (MIMO) ANTENNA SYSTEM

The antenna is a major aspect in the communication sector and has changed as needed to build a communication link. Significant capacity and reliability of the communication link are essential in today's wireless communication world. This is feasible by employing multiple radiators on both the transmitter and radiator sides, a technique known as a MIMO antenna system is used [10-11]. Multiple antennas on both sides, namely, transmitter and receiver, produce severe mutual coupling, lowering the performance of the antenna system. As a result, it is difficult to design a MIMO antenna with effective control (i.e., coupling reduction in the target frequency range without affecting other parameters such as reflection coefficient, antenna gain, and so on).

The idea was first proposed around 1970, and the first applications were reported in the 1990s. High capacity is obtained in conventional communication systems with a single antenna at transmitter and one antenna at receiver by employing either a larger bandwidth or a high-power transmitter. The total performance of a typical communication system is fading. MIMO reduces fading and increases channel capacity without increasing transmission power or bandwidth. There has recently been tremendous advancement in this domain. The MIMO antenna has gained popularity in recent years as a result of the growth of microwave integrated circuit (MIC) and monolithic microwave integrated circuit (MMIC) technology [12].

Printed antenna structures have become the most prevalent way to create microwave and milli-metre wave communication systems throughout the decades. As a result of the desire for high data rates and minimal latency, wireless communication standards have progressed. According to Cisco's annual internet report, by 2023, fixed broadband speeds will have doubled, 70 percent of the worldwide population will have mobile access, and 1.5 billion people will have 5G enabled devices. Because of the increased demand for high data rates with enhanced quality of service (QoS), Multi-Input-Multi-Output (MIMO) technology has emerged. MIMO utilizes numerous antennas on both the transmitter and receiver sides to boost the overall effectiveness of the communication system [13].

The mutual coupling of the radiating elements inside the MIMO antenna design is a significant disadvantage of the MIMO system. This indicates that energy will be transferred from one element to the other and vice versa. The close proximity of several radiating elements within the MIMO antenna arrangement is the cause of the problem. As a result, if radiating elements are closer together, the coupling between them will be stronger. Mutual coupling between MIMO antenna elements, on the other hand, affects the orthogonality of separate data channels, lowering MIMO system channel capacity. Various decoupling techniques, such as electromagnetic band-gaps, neutralisation lines, parasitic decoupling elements, decoupling network, LC tanks with resonant slots, strips or lumped components, decoupling surface, and pattern or polarisation diversity techniques, have been widely researched in recent years to reduce unnecessary mutual coupling.

MIMO antenna is one of the primary reasons for the availability of 3G and 4G wireless systems. Massive MIMO will be employed in future 5G systems. The usage of MIMO systems increases data rates, and the block diagram for a MIMO system is given in Figure 1.2. A MIMO system is made up of three basic constituents: the transmitter, the receiver, and the channel. A MIMO system, as opposed to a normal single antenna system, has multiple transmission M and multiple receiving antennas N. Each receiving antenna can receive signals from all emitting antennas, resulting in a total of M X N signal transmission channels.

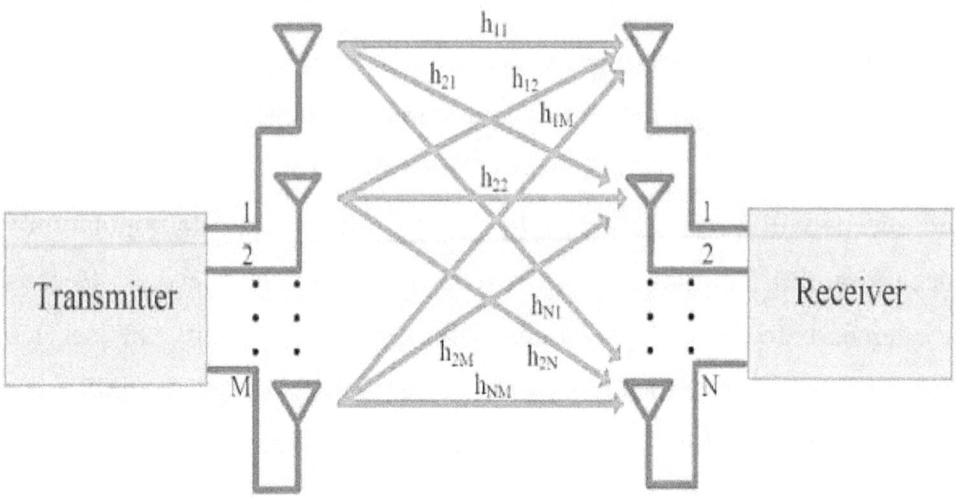

Figure 1.2. Block Diagram of a MIMO Antenna

Multiple antennas have not been utilised in wireless communication for the initial moment. Typically, several antennas (for e.g., antenna arrays) were employed to improve system SNR through beamforming, which focused antenna incident radiation in a definite direction. Later, multi-antennas were employed to provide spatial variety, counteracting the effects of multipath fading. Signals from distinct propagation pathways can be collected sequentially by multiple receiving antennas at a certain distance apart and then combined together constructively by the multiple antennas. Because the potential of all signal routes experiencing strong multipath fading is much smaller, the system Bit Error Rate (BER) is significantly decreased in this manner.

However, the benefits of spatial diversity from many antennas are constrained. The additional spatial multiplexing provided by multiple antenna systems is what makes it strong and desirable. Multiple independent data streams can be transferred via multiple independent signal propagation routes between the multiple sending and multiple receiving antennas. Consider the illustration in Figure 1.2. Considering no signal and noise correlation at both the transmitter and receiver ends, the MIMO system's capacity rises linearly with respect to the minimum number of transmitting/receiving antennas (min {M, N}), at no additional cost in terms of power or spectral response. MIMO has become a key component of several wireless communication technologies, including 4G LTE, WiMAX, and 5G, because to its huge advantage in spectral efficiency over conventional single-antenna systems.

1.3.1 Traditional MIMO and Massive MIMO

A Comparative study of typical MIMO for 4G Communication and the massive MIMO used in modern 5G technology is summarised in Table 1.1 below:

Table 1.1. Table explaining the features of traditional and massive MIMO technologies

Parameter	Traditional MIMO	Massive MIMO
Number of antenna elements	Up to 8	More than 100
Channel matrix requirement	Low	High
Channel capacity	Low	High
Noise resistance	Low	High
Diversity gain	Low	High
Correlation among antenna elements	Low	High
Symbol error rate performance	High	Low

Some of the technical advantages of Massive MIMO is that:

i. Reduced Power Consumption.

ii. Rise in the number of antennas reduces the thermal noise effects and small-scale fading.

iii. Improved Spatial Resolution

Because of the numerous benefits of massive MIMO technology, it will play an important part in fifth generation (5G) applications. In besides the advantages described above, it can significantly boost system throughput [14].

1.3.2 Millimeter-waves and Massive MIMO

With the rise of applications for mm-wave frequencies, massive MIMO technology consists of the employment of a large number of antenna array elements for wider applications such as beam forming/beam steering and spatial multiplexing for multiple user systems [15]. An antenna system that promotes transmission within the higher frequency band while ensuring optimum energy and spectral efficiency is necessary for mm-wave propagation. The need is met by using Massive MIMO at base stations [16]. As opposed to the sub-6 GHz frequency range, the mm-wave frequency band provides broader bandwidths with extremely high gains. However, the influence of noise power is greater in higher frequency bands. As a result, high gain beam steerable antennas with directional patterns are important for achieving high SNR [17]. In comparison to contemporary wireless communication technologies, mm-wave technology offers a variety of advantages, including larger bandwidths and higher capacity, lower intercepting probabilities, antennas with very small dimensions, and so on [18].

1.3.3 Sub-6 GHz and Massive MIMO

The mm-wave spectrum appears to be required to achieve the desired 5G data rates of the order of 20 Gbps. However, there are a number of challenges that must overcome before mm-wave can be used for mobile communications. The original equipment manufacturers are constantly developing mm-wave technology; nonetheless, in the near future Sub-6 GHz will be the critical spectrum for 5G networks in the short term. With the potential to carry large data rates over extended distances in both rural and urban regions, sub-6 GHz frequencies will play an essential role. Sub-6 GHz huge MIMO will significantly reduce interference difficulties. This can be performed by expanding the number of antennas in base stations, providing them to serve the vast majority of large urban users [17].

1.3.4 Challenges for antenna design in MIMO technology

The antenna designer sector confronts a challenging problem in designing antennas for wireless mobile communication systems. To prolong the operation of mobile services and to decrease system complexity, 5G communication systems require broad band and multi band antennas. Furthermore, the antennas encounter difficulties such as reduced physical dimensions, efficiency of integration with handheld devices, multi-antenna system operation, and effective MIMO operation. The fundamental procedures for constructing a MIMO antenna system are often divided into two categories.

1. Antenna design and implementation
2. Performance optimization analysis

The first stage in developing a MIMO is determining the appropriate operating frequency. The number of antenna elements used in the design has a large impact on the system's capacity and gain. As a result, it is critical to consider the trade-off between these used qualities. Another significant issue to consider while designing is the estimation of channel capacity and the reduction of mutual coupling, which increases port isolation between antenna parts. More parts are necessary to boost the system's throughput, which also increases the bandwidth of the system architecture.

1.3.5 Classification of 5G MIMO Antennas

The number of antennas chosen is an important criterion for classifying 5G MIMO designs. Table 1.2 illustrates the most commonly utilized antenna configurations suitable for 5G applications. It also has some major benefits and drawbacks of various types of antennas suitable for 5G systems. As illustrated in Figure 1.3, 5G antennas can also be categorised based on the ports used, namely antennas with single input and output ports and antennas with multiple input and output ports. The 5G antenna designs based on the number of I/O (input-output) ports are briefly defined as follows:

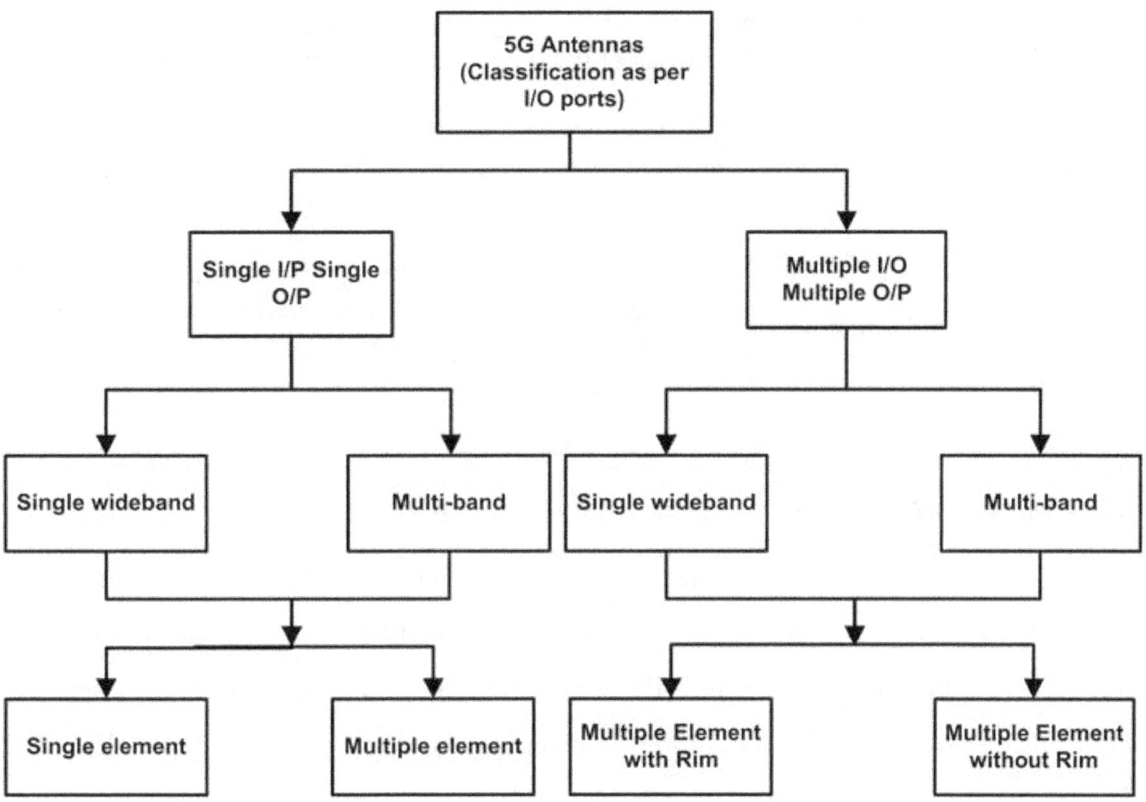

Figure 1.3. Organization of MIMO antennas related to I/O Ports

Table 1.2. Table explaining the features of the different antenna types used for 5G applications

MIMO Antenna Types	Explanations
Monopole	It is essentially a "λ/4" micro-strip line, but several variants have been documented, including helical, conical, and L-shaped patterns portrayed in compliance with the application and the MIMO requirements [18].
Dipole	It is a monopole antenna outgrowth consisting of two "λ/4" strip lines fed in between them. A variety of dipole radiating elements are defined for 5G purposes [19-20].
Magneto-Electric Dipole	The antenna has two kinds of dipoles: planar electric and vertical magnetic shorted. The feed is situated on the substrate's bottom side and stimulates the magnetic dipole initially [21-23].
Fractal	The same pattern is performed a number of times according to the mathematical concept of iteration. The patches are available in a range of shapes [24-25].
Inverted-F antenna	An antenna made of micro-strip lines bent with respect to one vertical line to produce an inverted F-shape. The vertical line is primarily fed by coaxial feed [26].
Planar Inverted-F antenna	It is a type of small patch antenna with two planes connected by shorting vias and supplied primarily from the bottom part [27].
Antipodal Vivaldi antenna	Two similar conductors are inserted at the top and bottom of the substrate, with the conductor on top serving as a condenser and the component on the bottom serving as the ground [28-32].

1.3.6 Evolution of Spectrum towards 5G Communication Systems

The key goal of future research will be to use 5G available spectrum alongside all previously operating cellular frequency bands. The majority of economies have intensified their research into 5G technologies. The sub-6 GHz and mm-wave band allocations for 5G spectrum are included. The sub-6 GHz bands primarily include n77 (3.3-4.2) GHz, n78 (3.3-3.8) GHz, n79 (4.4-5) GHz, and LTE-U (5.15-5.925) GHz; however, the majority of the effort is concentrated on (3.4-3.6) GHz and (5.15-5.925) GHz bands. (24.25-39.5) GHz, (37-43.5) GHz, (45.5-47) GHz, (47.2-48.2) GHz, and (66-71) GHz are the milli-meter-wave bands. In practice, equipment that operates in both high- and low-frequency bands gives significantly faster data speeds of around 20 Gbps and more than 100 Mbps. 5G is defined by low latency, high mobility, and the capacity to handle more users at once with higher antenna strengths, larger bandwidths, and lower radiation losses.

1.3.7 Challenges in the design of MIMO for 5G Applications

A few notable problems in the antenna design must be addressed in order to permit key trends. The section covers some of the significant issues that must be addressed.

a. 5G is intended to enable a wide range of client scenarios, including massive IoT, specific control missions, and improved mobile broadband. 5G should be capable of functioning on a number of bands and spectrums, including shared bands, approved bands, unauthorized bands, and lower sub-6 GHz to millimeter-wave bands, to ensure spectrum availability. Providing the most effective functionality to meet spectrum and user case requirements is a big challenge for 5G antenna designs.

b. Another issue for 5G antenna design is to provide for continual technical advancement in addition to forward compatibility. The primary components to be examined for wireless applications are as follows: Spectrum flexibility is required to support devices operating in different spectral bands. To accommodate the large range of possible end destinations, an air interface that is scalable is necessary. The adaptability of air interfaces in the future is crucial for technological advancement and addressing future demands.

c. With the expansion of IoT devices in the future years, there will be an unavoidable necessity for huge connection for wireless systems. This ever-increasing need for machine-to-machine connectivity is defined by excellent reliability, reduced transmission intricacy, very low power consumption, minimal latency, and so on. These issues must be overcome, and 5G networks are expected to be more network-centric, less affordable, and primarily targeted toward human-based applications.

1.4 APPLICATIONS OF MIMO SYSTEMS

1.4.1 Wireless Communication

MIMO serves as the backbone in modern wireless protocols such as 4G Long Term Evolution (LTE), LTE advanced, and IEEE 802.16 Worldwide Interoperability for Microwave Access (Wi-MAX). It can accomplish reliability and transmit diversity through the use of Single-User MIMO, Multi-User MIMO and beamforming. SU-MIMO is implemented for downlink in LTE, with two or four transmit antennas used to send multiple data lines to a single user device, and two antennas used for uplink [33-34]. Wi-MAX achieves spatial multiplexing and transmit diversity by employing two to four antennas at the transmitter. The High Throughput Wireless LAN (WLAN) standard 802.11n, launched in 2009, employs techniques such as multiple spatial data streams using 4x4 MIMO antennas with beamforming techniques to reach data rates of up to 600 Mb/s with four spatial data streams of 40 MHz bandwidth. MIMO technology is an important enabler of fifth generation (5G) wireless communication systems. It is intended to provide peak data rates of up to 20 Gbps by leveraging mm-Wave and sub-6 GHz frequencies.

1.4.2 RADAR Technology

The usage of MIMO in radar technology is a new field of research that has piqued the interest of investigators [35-36]. Target RCS (Radar cross section) is scintillated in a MIMO radar system by broadcasting several signals from substantially uncorrelated transmit antennas. Incoming signals are merged at the receiver to achieve high SNR compared to traditional radar. The received signal power varies with conventional radar [37-38].

Due to the different waveforms, the reverberation signals can be redistributed to the same transmitter. The conceptual outcome of an antenna field of N transmitters and a field of K receivers is a virtual field of K-N elements with an expanded virtual aperture. MIMO radar techniques can increase spatial resolution and provide much improved resistance to interference. The possibility of finding targets is boosted by increasing the signal-to-noise ratio.

1.4.3 Vehicular Applications

MIMO has been utilised to boost data throughput and accuracy in Unmanned Aerial Vehicle (UAV) applications, where MIMO antenna location and orientation on the UAV are critical design constraints. Multi-user MIMO has been employed in UAV communication applications to simultaneously provide numerous ground terminals [39-40]. Beamforming has also been used to enhance data communication. MIMO has been utilised in automotive wireless communications, in which antennas positioned on automobiles connect with the driver and passengers, roadside base stations, satellites, and nearby vehicles. Vehicle base station antennas are used to offer consistent signal quality in covered areas [41].

1.4.4 5G Applications

Applications utilising Sub-6 GHz 5G bands are becoming increasingly significant as the demand for expanded services and better speeds grows. Lower bands below 6 GHz meet current data throughput requirements in Gbps and provide a greater coverage area. The advancement of technology creates issues in system design, as connected smart devices require the integration of antennas and RF circuits operating in 5G bands. As 5G deployment accelerates, there is a greater need for novel antenna designs that can meet the needs of users.

1.5 ANTENNA DIVERSITY TECHNIQUES

Frequency, time and antenna diversity given in Figure 1.4 are commonly employed to decrease spatial correlation between antennas and to make replication of a signal for diversity reasons. When used in MIMO systems, all multi-element antennas use these diversity strategies to produce great performance.

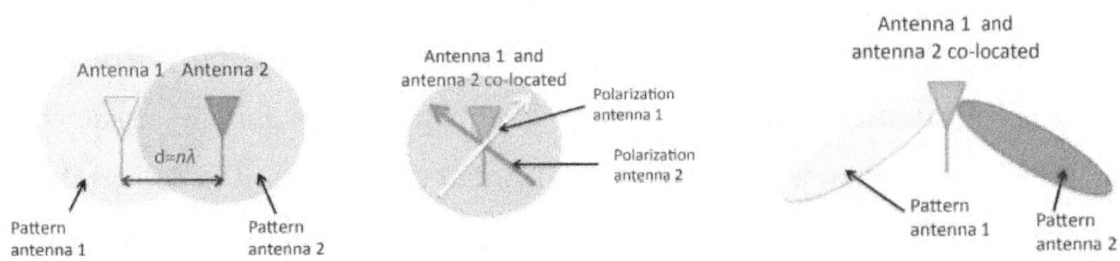

Figure 1.4. Antenna Diversity Techniques

1.5.1 Spatial Diversity

Spatial diversity, as shown in Figure 1.5, is the easiest and most prevalent type of diversification in wireless communications. The phase delay among antennas at two different sites in space allows the fading signal in the second antenna to be mutually independent with the first antenna. To achieve adequate decorrelation, the antennas must be spaced at a minimal level depending on the surrounding environment and, in general, the width of the multipath angle of arrival. It has been demonstrated that in an urban region, a spacing higher than 10 λ is required for effective decorrelation, however in an interior environment with substantial scattering, a minimum distance of only 0.5 λ is required. The figure depicts two base station antennas that make use of spatial diversity.

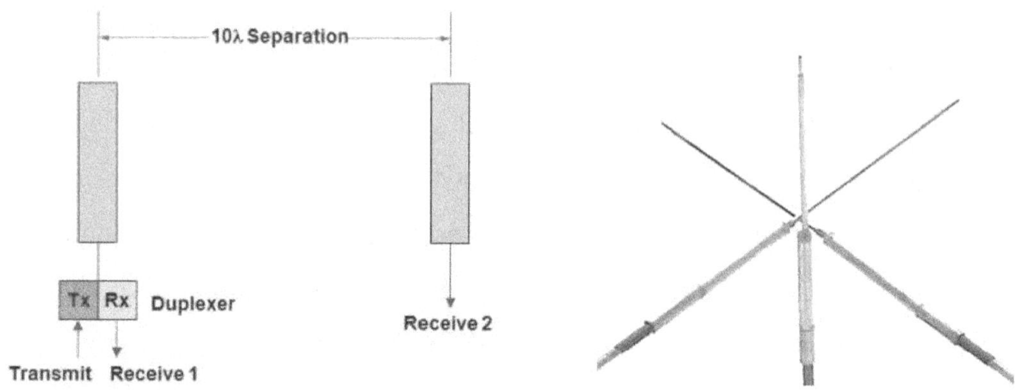

Figure 1.5. Spatial and Polarization Diversity

1.5.2 Polarization Diversity

In contrast to other diversities, the advantages of polarization diversity include the possibility of locating the antennas in the same location. The effect of polarisation diversity on sub-channel decorrelation and channel conditions in MIMO systems has been examined using simulation and observations, mostly using dual-polarized antennas. It has been demonstrated that a dual-polarized system of co-located antennas can provide nearly the very same channel capacity as spatial diversity. Even when it was initially assumed that only two different polarizations (vertical and horizontal) could be used to de-correlate the data transmission sub-channels, it has now been demonstrated that it is possible to use all three electric field and magnetic field components, as shown in Figure 1.5, to significantly increase the capacity of a MIMO communication link.

1.5.3 Radiation Pattern Diversity

If several antennas have distinctive patterns, the diversity gain can be significant. The antenna system in pattern diversity includes two or more co-located antennas with distinct emission patterns. This sort of diversity employs directional antennas that are often separated physically by some short distance. They can differentiate a substantial amount of the angle space and provide a higher gain than a single omnidirectional radiator. This type of diversification can ensure the reception of several sets of multipath waves. Pattern diversity was utilised in base stations, and it performed similarly to spatial diversity, especially in congested metropolitan regions. The below Figure 1.6 is a 5G base station setup that makes use of pattern diversity.

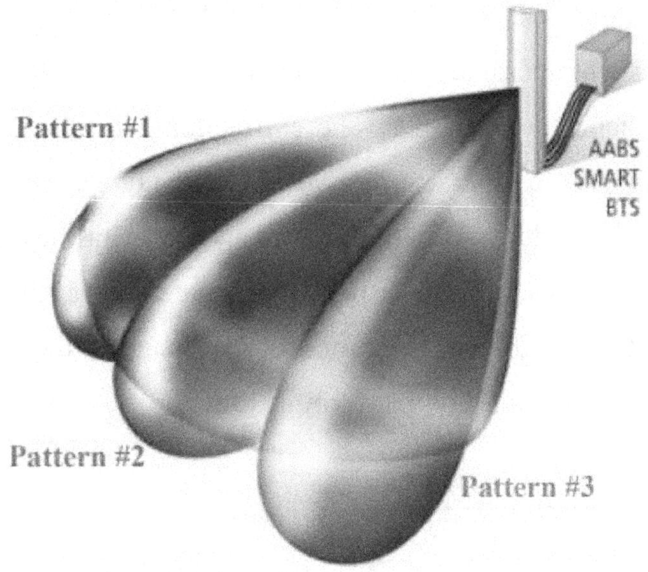

Figure 1.6. Pattern Diversity

1.6 MUTUAL COUPLING AND MIMO ANTENNA PERFORMANCE PARAMETERS

The designs of MIMO antennas for mobile and base station applications, as well as decoupling approaches, are described. Techniques for improving isolation include the use of neutralisation lines, slotted and defective ground structures, parasitic elements, ground strips, meta-materials etc. The approaches used are determined by the design of MIMO antenna elements as well as their frequency of operation. The relationship between the antenna elements is also discussed, as well as their impact on antenna performance.

According to modern technologies for 5G systems, adopting MIMO antenna systems is one of the critical components that must be handled and carefully designed. Fundamental performance metrics such as impedance bandwidth, resonance frequency, gain, antenna efficiency and radiation pattern are used to evaluate single antenna designs. However, in the scenario of MIMO antennas with many antenna elements, several additional performance metrics are required for a comprehensive MIMO antenna system characterization. The subsections that follow will explain these MIMO performance measures.

1.6.1 Envelop Correlation Coefficient (ECC)

ECC (Envelop correlation coefficient) assesses the consequences of MIMO system performance deterioration caused by mutual coupling across antenna elements. In which, it depicts antenna element isolation as a result of radiation patterns. When all ports are activated or performed at the same time, the radiation pattern of all elements may have an influence on the system's performance. The correlation coefficient indicates how much communication channels are separated or associated with one another. This statistic takes into account the antenna system's radiation pattern and how significantly the patterns affect one another when used concurrently (especially in the case of MIMO antenna system). The envelop correlation coefficient is defined as the square of the correlation coefficient [42]. It is defined by using the following formula as:

$$\rho_e = \frac{\left|\iint_{4\pi}\left[\vec{F_1}(\theta,\varphi) * \vec{F_2}(\theta,\varphi)\right]d\Omega\right|^2}{\iint_{4\pi}\left|\vec{F_1(\theta,\varphi)}\right|^2 d\Omega \iint_{4\pi}\left|\vec{F_2(\theta,\varphi)}\right|^2 d\Omega} \tag{1.1}$$

Where $\overrightarrow{F_i(\theta,\varphi)}$ is the antenna's three-dimensional field radiation pattern when the ith port is activated, and is the solid angle. The Hermitian product operator is symbolized by (*). Equation (1.1) is a challenging formula that requires three-dimensional radiation pattern observations as well as numerical methods. When examining a homogenous multipath environment with balanced polarisation, this is valid. When examining a unidirectional and single mode antenna, a simple derivation demonstrates that the correlation coefficient can be calculated using the reflection coefficient. In the equation (1.2) below, the function of performance on the calculation of the envelope correlation coefficient from the S-parameters was added.

$$|\rho_{ij}| = \rho_{eij} = \left| \frac{|S_{ii}^{*}S_{ij} + S_{ji}^{*}S_{jj}|}{\left|\left(1-|S_{ii}|^{2}-|S_{ji}|^{2}\right)\left(1-|S_{jj}|^{2}-|S_{ij}|^{2}\right)\eta_{radi}\eta_{radj}\right|^{0.5}} \right|^{2} \quad (1.2)$$

Where ρ_{ij} is the correlation coefficient between elements *i* and *j*, ρ_{eij} denotes the envelop correlation coefficient, S_{ij} denotes the s-parameter between the *i* and *j* elements (coupling), and η_{radi}, η_{radj} denotes the radiation efficiency of elements *i* and *j*, respectively [43]. In this calculation, just the S- parameters and radiation efficiencies that can be easily evaluated in contrast to the requisite 3D radiation patterns are essential. A correlation coefficient of 0.3 has been determined as an appropriate value for 4G wireless applications. It has been shown that under-estimates the correlation coefficient values of a MIMO antenna system and presents an unduly optimistic perspective, whilst the reality is significantly worse. Since it is based on authentic radiated fields that directly affect the channels, it provides a more precise measure of the correlation coefficient and should be used all of the time. Returning to a dual-element printed antenna system, the considered envelope correlation coefficient from the radiated fields was 0.29 and 0.22 at the lower and higher bands, respectively, while utilising the S-parameters it was 0.03 and 0.009 at the lower and higher bands. The distinction between the two approaches is evident [44].

1.6.2 Total Active Reflection Coefficient (TARC)

The reflection coefficient is used to compute the bandwidth and gain for a single antenna. However, in a multi antenna and/or multiport design, this measure may not be appropriate for evaluating the impedance bandwidth and efficiency of the MIMO antenna system. The total active reflection coefficient given in Equation (1.3) in a multiport system provides information about the radiated power and effective operational bandwidth of multiantenna systems [45]. The square root of total reflected power divided by total incident power is described as TARC. TARC values ranging from 0 to 1. In this situation, a number of 0 indicates that all power is radiated, whereas a value of 1 indicates that all power is reflected. TARC can be calculated using the MIMO antenna system's S-parameters. It is given for an N-element antenna by,

$$\Gamma_a^t = \frac{\sqrt{\sum_{i=1}^{N}|b_i|^2}}{\sqrt{\sum_{i=1}^{N}|a_i|^2}} \tag{1.3}$$

where a_i and b_i represent the incident and reflected signals, respectively. TARC accepts values between 0 and 1, with zero indicating that all powers were radiated and one indicating that all incident power was reflected but no power was radiated [46]. The summation of the powers available on all of the antenna system's ports is the total power. It also takes into account the impact of feeding phase changes between antenna ports. As a result, for a given phase excitation between its ports, a single TARC curve can be utilized to calculate the frequency response and impedance bandwidth of the overall antenna system. For a 2-port MIMO system, it can be calculated by using the equation (1.4) as follows:

$$\Gamma_a^t = \frac{\sqrt{\left(|S_{11}+S_{12}\,e^{j\theta}|^2\right)+\left(|S_{22}+S_{21}\,e^{j\theta}|^2\right)}}{\sqrt{2}} \tag{1.4}$$

where θ is the input feeding phase, S_{xx} is the port's reflection co-efficient, and S_{xy} is the coupling between the two ports connected to the antenna structure. After determining the S-parameters of a two-port network, the irregular phase is drifted between 0° and 180° to examine the impact of phase variation between the two ports on the resonance conduct of the antenna, and the corresponding TARC curves for impactful bandwidth assessment are generated.

1.6.3 Mean Effective Gain (MEG)

For the MIMO antenna system, the conventional gain of an element antenna is not the accurate method. Thus, in a MIMO system, the parameter active gain is employed, which is the gain associated with an excited port while all other ports are ended with matching load. However, in MIMO, one must also consider the impacts of various propagation circumstances [47]. As a result, Mean Effective Gain (MEG) is utilized to evaluate the diversity performance of a MIMO system in a variety of propagation environments by combining the effects of active antenna gain with numerical propagation characteristics of input signal [48].

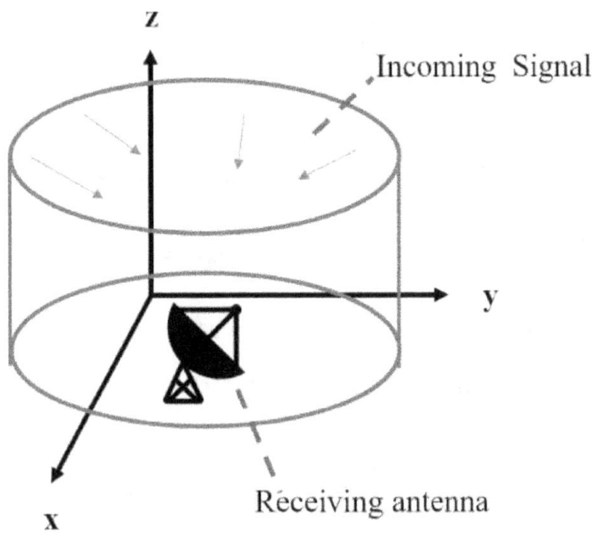

Figure 1.7. A representation of incoming signal in the XY plane

The transmitted signal can enter at a receiver positioned in a three-dimensional plane, as shown in Figure 1.7 as an incoming signal from any direction in the XY plane. As a result, the overall average received power will be affected by the receiving antenna's gain pattern as well as the statistical angular distribution (probabilistic model). The MEG is the ratio of mean received power to mean incident power, and for best performance, the ratio of MEG_i (i^{th} antenna) to MEG_j (j^{th} antenna) should approach unity or 0 dB. As a result, the Mean Effective Gain (MEG) given in equation (1.5) is commonly used to compensate for all of the impacts of the Total antenna efficiency, Antenna gain, and Wireless environment. The formula is as follows:

$$MEG = \oint \left(\frac{XPR}{XPR+1} G_\theta(\Omega). P_\theta(\Omega) + \frac{1}{XPR+1} G_\Phi(\Omega). P_\Phi(\Omega)\, d\Omega\right) \quad (1.5)$$

Where, $d\Omega = \sin\theta\, d\theta\, d\Phi$, XPR is the cross-polarization ratio, $G_\theta(\Omega)$ and $G_\phi(\Omega)$ are the θ and ϕ components of the realized antenna gain pattern, respectively. The antenna realized gains in MEG are normalized to the given equation (1.6):

$$\oint \left(G_\theta(\Omega) + G_\Phi(\Omega)\right) d\Omega = 4\pi \quad (1.6)$$

The statistical power spectrum distributions of both vertically and horizontally polarized incident radio waves can be represented by P_θ (Ω) and (Ω), respectively. If P_θ (Ω) and (Ω) are separable in elevation and azimuth, they can be described by the following Equations (1.7 & 1.8):

$$P_\theta(\Omega) = P_\theta(\theta,\Phi) = P_\theta(\theta)P_\theta(\Phi) \tag{1.7}$$

$$P_\Phi(\Omega) = P_\Phi(\theta,\Phi) = P_\Phi(\theta)P_\Phi(\Phi) \tag{1.8}$$

$P_\theta(\Omega)$ and $P_\Phi(\Omega)$ are standardized and given in the equation (1.9):

$$\oint(P_\theta(\Omega))\, d\Omega = \oint(P_\Phi(\Omega))\, d\Omega = 1 \tag{1.9}$$

XPR (cross polar ratio) is the ratio of time average vertical $P_\theta(\Omega)$ power to time average horizontal (Ω) power given in the equation (1.10).

$$XPR = \frac{P_\theta(\Omega)}{P_\Phi(\Omega)} \tag{1.10}$$

In an isotropic environment, it can be characterized by XPR=1, $P_\theta(\Omega) = (\Omega) = 1/4\pi$, and $MEG = \eta/2$, where η is the total antenna efficiency. In order to get an optimal diversity performance, the multiple antenna system has to satisfy the balance power requirement in the given equation (1.11).

$$MEG_{antenna\ 1} \approx MEG_{antenna\ 2} \tag{1.11}$$

1.6.4 Diversity Gain

The Diversity Gain (DG) seems to be an another important diversity feature that may be determined using the below Equations due to the enhanced possible DG and correlation coefficient. Diversity gain refers to any diversity strategy that optimises the signal-to-interference ratio. Because it has a larger distribution gain value, the antenna provides superior isolation, and vice versa. The equations explain the representations of the current antenna's diversity gain and indicate that the DG values for both cases are about 10 dB at the functional band [49].

$$ADG = 10\sqrt{1 - |ECC|}^2 \tag{1.12}$$

$$EDG = \eta_{total}\ X\ ADG \tag{1.13}$$

ECC (m, n) and *DG* (m, n) characteristics are same with *ECC* (n, m) and *DG* (n, m) when m not equal to n since MIMO antenna elements are similar with even spacing. *ECC* values in the operational bands are observed to be less than 0.1, and *DG* is extremely close to 10, indicating that the proposed MIMO antenna has successful diversity characteristics. The SNR after implementing the MIMO system is effectively increased as measured by the ADG. The acceptable maximum ADG level is 10 dB. The *EDG* is another practical *DG* variant that accounts for antenna radiation losses. The *ECC* values calculated above are used to calculate the *ADG* and *EDG* using formulas (1.12) and (1.13). For a MIMO system to work effectively, both the *ADG* and *EDG* must be higher than 7.5 dB.

1.7 SIMULATION TOOL

The implementation of effective simulation software is critical to the successful attainment of our objectives. The software utilised should make it easier to calculate the positioning of the antenna's feed point and optimise the various parametric dimensions, as well as the structure's theoretical behaviour. Prior to construction and measurement, the antennas reported in this thesis are designed and parametrically studied using the commercial software Ansys High Frequency Structure Simulator (HFSS) and CST Microwave Studio.

1.7.1 Ansys HFSS

HFSS is a high-performance full-wave electromagnetic (EM) field simulator based on the Finite Element Method for arbitrary 3D volumetric device design. Simulation, modelling, visualisation, and automation have all been merged into a user-friendly environment. HFSS was used to optimise the structure's position and size, as well as to study their parametric variation. The definition of the system's geometry is the first step in the process of replicating a structure in HFSS. A significant portion is assigning the material properties and 3D or 2D border elements are visible in the HFSS windows. Wave port, lumped port incident wave scheme, and so on are used to select the proper port excitation. Around the system, an appropriate volume of air-filled radiation boundary is created. The frequency sweep range and number of frequency points are then selected. The simulation engine is now operational. Following the completion of the simulation, the results, such as scattering parameters, radiation patterns, and field distributions, can be exhibited. The HFSS package has optimization tools that can assist the designer in optimising the settings and performance. The scalar and vector representations of the E, H, and J fields help to clarify the structure under investigation. The HFSS package has optimization tools that can assist the designer in optimising the settings and performance. The scalar and vector representations of the E, H, and J fields help to clarify the construct under investigation.

1.7.2 CST Microwave Studio®

CST Microwave Studio® (CST MWS) is a cutting-edge simulation tool for high-frequency (HF) devices such as antennas, filters, couplers, planar and multi-layer structures, and SI and EMC effects. In CST, the antenna structure is designed by defining the coordinates for each point on the structure. Dielectrics and metallization are assigned when the structure is created. To account for losses appropriately, metallization is assumed to be copper. In all simulations, a finite metal thickness of 10 μm is specified. CST provides problem-specific templates with pre-defined or user-defined characteristics such as initial conditions, mesh size, type, and so on. Then, to provide solutions for all desired frequencies, a frequency sweep is created, and a waveguide port is designated to stimulate the antennas. Antennas are modelled in this way to analyse the return loss, radiation pattern, gain etc.

1.8 ANTENNA FABRICATION AND EXPERIMENTAL ANALYSIS

This section provides a quick overview of the steps involved in the fabrication of the various antenna prototypes.

1.8.1 Substrate Selection

The primary and most important step in the fabrication process is to choose a suitable substrate. There is no ultimate substrate, but it should be chosen so that its electrical and material qualities match the application's requirements. Material properties such as the dielectric constant and loss tangent are crucial. Because the feed is a coaxial probe, a double-sided substrate is necessary for this task [50-54]. Polytetrafluroethylene (PTFE), polystyrene, polyolefin, polyphenylene, alumina, sapphire, quartz, ferromagnetic, rutile, RT Duroid, and semiconductor substrates are examples of commercially accessible substrates. Despite their high radiation efficiency and electrical performance, PTFE and quartz are usually costlier substrate materials. Other flexible substrate materials are readily available, permitting the antenna to be placed on curved surfaces. Surface waves will be excited in high dielectric constant substrates, resulting in unwanted radiation from the antenna in undesirable directions. To reduce the propagation delay and to maximize the signal speed, a low relative permittivity with a low dielectric loss material is required. Furthermore, the material should have a good thermal conductivity to dissipate heat. Thickness, uniformity, isotropicity, and dimensional strength are also key substrate characteristics. In this work, FR4, a low-cost, widely available, and frequently used substrate material for microwave applications, was used for easy antenna fabrication [55-56].

1.8.2 Fabrication Process

Considering the narrow bandwidths of microstrip antennas, structural constraints in their fabrication are critical. As a result, tremendous consideration must be taken during the fabrication process. The typical steps in the fabrication of a microstrip antenna are as follows: Creating artwork from designs is the first step in the fabrication process. It is then scaled on a manually operated coordinagraph with a precise cutting blade. At this step, quality is crucial, and depending on the antenna's complexity and dimensions, fully or increased scale artwork should be created on film. The upgraded artwork should be photo reduced with a high precision camera to produce a high-resolution negative, which will then be used to reveal the photo resist. The laminate should be cleansed to guarantee optimal photo resist adhesion and resolution during the photo developing process. Using a laminator, apply the photo resist to both sides of the laminate. Laminating a metallic foil to a dielectric substrate sometimes requires raising the temperature so that the laminate can cool to ambient temperature before exposure and development.

To accomplish the desired fine-line resolution, a vacuum frame copy board or other techniques is used to keep the photographic negative in close contact with the polyethylene cover sheet of the attached photo resist. The outer photo resist polymerizes when exposed to the correct wavelength of light, making it insoluble in the developer solution. The backside of the antenna is completely visible without a mask since the copper foil is kept to act as a ground plane. The protective polyethylene cover sheet of the photo resist is detached, and the antenna is now developed in a developer, which dissolves the soluble photo resist component. To ensure correct development, visual inspection is used. The following process is etching, which is crucial and requires great attention. After etching, the processed photo resist is removed with a stripping solution. To ensure a high-quality product, visual and optical testing are performed. Thermal cover bonding can be used if required. The antenna's edges are smoothed, then it is cleaned and dried.

The process outlined above is commonly used to create commercial micro strip antennas. Due to the time constraints, the approaches adopted during the project to develop the antenna is a rapid fabrication process. A square of size λ_o is cut from a single clad FRP dielectric substrate. The patch is made of self-adhesive copper sheet that has been trimmed to size. After that, the patch is applied on the laminate. The coaxial feed hole is then located using a drill bit. After that, a connector is attached to the back of the antenna. As a result, the antenna has been completed and is ready for testing.

1.8.3 Photolithography

Photolithography, sometimes referred to as optical lithography, is the method of applying geometric shapes from a photo-mask to the surface of a substrate. The selected single-side substrate should be produced with the ideal antenna design. Any CAD programme is used to carefully create the layout. This is printed on butter paper or clear OHP paper. An inverse of the design is used to construct the mask since a negative photo resist is used in photolithography. A metal cloth is used to remove any flaws from the substrate that has copper on one side, and acetone is used to clean the metal surface. Particularly when the operating frequency is very high, any variation in the etched structure will result in a departure from the predicted parameters for the resonant frequency. A small layer of photo resist, mixed with thinner in a 1:2 ratio, is applied to the substrate's surface and allowed to dry. The photo resist is then covered with the mask, which is subsequently exposed to UV light for a predetermined period of time. The photo resist hardens at this time as the UV light is directed at the locations where the substance is required. The developer solution then gets rid of the resist in the parts that weren't exposed. To make the created structure readily apparent, the substrate is then coloured. After that, it is rinsed with running water. By etching it in ferric chloride, the undesirable metallic structure can be eliminated. Except for behind the solidified photo resist layer, $FeCl_3$ dissolves the copper components. The laminate is then gently washed with acetone solution to remove the solidified photo resist.

1.8.4 Antenna Measurement Facilities

The Anritsu Vector Master MS2038C Vector Network Analyzer, PNA E8362B Analyzer, and allied setup are used to test antenna parameters such as reflection and radiation. All measurements are automated and synchronised. This part provides a quick overview of the basic facilities used for antenna measurements [57].

1.8.4.1 Anritsu Vector Master MS2038C

Anritsu's MS2038C is a handheld combined analyzer that includes a vector network analyzer with a frequency range of 5 KHz to 20 GHz and a spectrum analyzer with a frequency range of 9 KHz to 20 GHz. It has a handheld VNA with a 12-term error correction algorithm and a typical measurement speed of 350 microsec per point. It can measure and display all four S-parameters simultaneously using a three-receiver topology and a single connection to the DUT. The MS2038C is appropriate for tuning filters in the field, where several S-parameters frequently interact.

1.8.4.2 PNA E8362B Analyzer

The Agilent E8362B PNA (Performance Network Analyzer) offers high performance, extensive automation options, flexible connectivity, and is simple to operate. Due to its quick sweep speed, broad dynamic range, minimal trace noise, and numerous connecting options, designers and engineers like the Agilent E8362B for testing excellent performance components.

Some of the other features and specifications includes:

- The operation range is from 10 MHz to 20 GHz
- 26 μs/point measurement speed, 32 channels, 16,001 points
- TRL/LRM calibration, on-wafer, in-fixture, waveguide, and antenna measurements
- Mixer conversion loss, return loss, isolation, and absolute group delay
- Amplifier gain compression, harmonic, IMD, and pulsed-RF

The measurement process is significantly made simpler by the Windows operating system and mouse interface. Through tests involving cavity perturbation and other reflection properties, this equipment is used in the thesis to calculate the dielectric constant.

1.8.4.3 Anechoic Chamber

An anechoic chamber is used to create the free space environment required for antenna pattern measurements. The anechoic chamber creates a quiet zone free of all Electro-Magnetic interferences. To prevent reflections from surrounding objects, all antenna characterizations are performed in an anechoic chamber. High-quality low foam that has been prepared with a dielectrically/magnetically lossy substance makes up the absorbers utilised to build the chamber [58]. Pyramidal and flat absorbers with polyurethrene foam bases and the proper sizes are utilised to cover the walls. The tapered pyramidal forms offer superior impedance matching for the microwave power that strikes it. To prevent extraneous interferences, the chamber is completely surrounded by an aluminium sheet on all sides. For a wide frequency range of 1GHz to 18 GHz, the polyurethrene foam structure enables geometrical impedance matching, while the scattered carbon provides the needed attenuation.

1.8.5 Measurement Process

The methodologies used to determine the antenna properties are detailed further below. In fact, the network analyzer is linked to huge cables and connectors. The connector and cables exhibit losses at higher frequencies, which should be avoided. To obtain accurate scattering parameters, the instrument should be calibrated with known standards of open, short, and matched loads. The network analyzer has numerous calibration techniques. Calibration methods such as single port, complete two port, and TRL are commonly utilised. The TRL calibration method can precisely measure the scattering parameters of two port passive or active devices. While calibrating, proper phase delay is introduced to ensure that the reference plane for all measurements in the intended band is truly at zero degree, thus accounting for possible cable length changes. This can be measured with the reflection coefficient, VSWR, input impedance, and other parameters using the single port calibration approach.

1.8.5.1 Measurement of S-Parameters, Resonant Frequency and Bandwidth

The Scattering parameter S11, also known as the Reflection coefficient (Γ), at the antenna input is the ratio of the reflected voltage or current to the incident voltage or current when the antenna is connected to network analyzer port 1. In terms of input VSWR or return loss, it reflects the degree of impedance mismatch between the antenna and the source line. The return loss (RL), denoted by RL = -20log (|Γ|) = -20log (|S11|) = -|S11| (dB), is the ratio of reflected to the incident power. The antenna is connected to any one of the network analyzer's terminals, and the VNA is set to measure the return loss characteristic in S11/S22 mode. The port is calibrated for the desired frequency range by using a short, open, and matched load. The AUT is now linked to the VNA's calibrated port. The magnitude and phase of the observed S11 versus frequency are then saved on the computer in comma separated variable (.csv) format using the automation software. While S11 or S22 represent the return loss at port 1 or port 2, S21 represents the transmission coefficient or the isolation between the antenna's ports 1 and 2. The frequency at which the curve exhibits the maximum dip (lowest value of return loss) is indeed the antenna's resonant frequency (f_r).

The frequency range for which the return loss value is within -10 dB points is commonly referred to as the antenna's 2:1 VSWR bandwidth. This is because VSWR = 2 corresponds to the reflection coefficient $\Gamma = \frac{VSWR+1}{VSWR-1} = \frac{1}{3}$ which gives RL = -20 log (1/3) = -10 dB. As a result, the 2:1 VSWR bandwidths at the two ports are estimated by monitoring the frequency range (Δfr) for which the return loss remains ≤ -10dB. The fractional bandwidth is calculated as $\frac{\Delta \mathrm{fr}}{\mathrm{fr}}$. Percentage bandwidth can be calculated by using the equation as

$$\frac{Bandwidth}{Centre\ frequency} \times 100. \tag{1.14}$$

The AUT's input impedance at resonant frequencies can be measured directly from the Smith Chart display in the network analyzer, where the centre corresponds to 50 Ω after calibration.

1.8.5.2 Radiation Pattern

The radiation pattern is determined using the setup within the anechoic chamber. The AUT, which is attached as a receiver to one of the VNA's ports, is placed on the automated turn table in the chamber's quiet zone. The transmitter is a conventional wideband horn antenna attached to the network analyzer's other port. The turn table is controlled by a STIC positioner controller. The AUT's primary E- and H-plane radiation patterns (with both co and cross polar patterns) are measured by keeping the network analyzer in S21/S12 mode and the frequency range within the -10 dB return loss bandwidth. By manually adjusting the positioner, the antennas are aligned at boresight for maximum reception. A THRU calibration is now done and saved in the CAL set for the frequency range of interest. Time domain gate parameters are supplied to avoid any artificial reflections that could distort the measured data [59].

1.8.5.3 Antenna Gain and Efficiency

When both thc antennas are given with the same input power, the gain of an antenna is defined as the ratio of the intensity of its radiation in the direction where it is greatest to that of a reference antenna. When an isotropic antenna is used as the reference, the gain is stated in decibels (dBi). An antenna's gain is a neutral phenomenon. The antenna does not increase power, but rather redistributes it to produce more radiated power in a specific direction than an isotropic antenna would. When the experimental setup is equal to that of the radiation pattern measurement, the gain comparison method is employed to estimate the absolute gain of the AUT. As the reference antenna, a conventional wideband horn antenna with a gain chart is employed. The antenna positioner is used to bore sight the reference antenna, which is then THRU calibrated for the frequency range of interest. After that, the horn antenna is substituted by an AUT aligned in the main beam direction, and the transmission coefficient S21 (dB) is measured. It represents the antenna's relative gain with respect to the reference antenna. This relative gain is multiplied by the actual gain of the standard antenna, as shown in the gain table provided by the manufacturer, to obtain the absolute gain. According to the IEEE definition, antenna efficiency is the ratio of total power radiated by the antenna to total power accepted by the antenna at its terminals during the radiation process [60-61].

1.8.5.4 Polarization

The experimental setup for measuring polarisation is identical to that of the radiation pattern. The AUT is configured in transmit mode, with the wide band horn serving as the receiver. The AUT is fed an RF signal, and the power received by the wide band horn in the vertical plane is recorded as a function of frequency. The transmitting antenna is then rotated through 90° and to measure the received power as a function of frequency in the horizontal plane. The measured data in both planes is processed to determine the polarisation of the test antenna over the whole operating spectrum.

1.9 OBJECTIVES AND SCOPE OF THE WORK

MIMO antennas are implemented in current telecommunication systems at several application levels, including LTE, Wi-Fi, WLAN, and 5G systems. It comprises of more than one monopole close, resulting in monopole mutual interaction. Mutual coupling between monopoles has a substantial impact on diversity performance. As a result, a compact MIMO antenna with strong isolation is sought for easy integration with external systems. The antenna uses diversity techniques to improve data rate, transmission capacity, and reliability while reducing bandwidth and power consumption. However, integrating many antennas in small terminals is difficult due to low isolation between closely located antennas, which reduces system performance. Mutual coupling can significantly degrade the adaptive antenna array's signal-to-noise ratio (SNR). It has the potential to degrade carrier frequency offset, channel estimations, and angle of arrival. The undesirable effect of low mutual coupling on a MIMO antenna's reflection coefficient should not be disregarded. As a result, the most difficult challenge is to build a small MIMO antenna with low mutual coupling. However, the majority of the papers published here provide minimal information of the systematic design approach as well as the numerous analytical techniques for improving MIMO system isolation. As a result, the proposed thesis focuses on small and high-performance MIMO antenna approaches with 5G Sub-6 GHz applications and recommends various associated solutions, including detailed simulation and fabrication work. For 5G systems, majority of research is going towards the milli-meter frequency range. While this range guarantees good bandwidth and provides the real estate for massive MIMOs it is still a challenge to integrate existing 3G/4G devices with 5G systems. In order to provide a viable solution to that, I have designed a MIMO antenna operating at two different frequencies.

The objectives and Scope of the present work is:

- To design a Novel MIMO antenna for 5G Sub-6 GHz applications.
- To design a Novel-shaped Hexa-Band Frequency Reconfigurable antenna integrated with PIN diode for S-Band and 5G Sub-6 GHz Applications.

- To design a MIMO antenna with (a) good Impedance Matching, S11< −10 dB (b) Mutual Coupling, S21< −20 dB (c) Antenna Efficiency > 70 % (d) Envelope Correlation Coefficient, ECC < 0.5 and (e) Diversity Gain (DG) > 9 dB.
- To employ various isolation techniques to reduce mutual coupling.
- To design compact MIMO antenna structures suitable for small mobile terminals. Since multiple antenna systems have large size, design of compact antenna is a demanding task.
- To analyse the reduction in mutual coupling of MIMO antennas using surface current distribution of antenna at various resonant frequencies.
- To evaluate the MIMO performance of antenna using metrics such as ECC and diversity gain.
- To compare the proposed work with the state-of-the-art techniques and similar antennas stated in the literature.

1.10 ORGANIZATION OF THESIS

The thesis is organized into seven chapters as follows.

Chapter I introduces the topic by explaining its evolution as well as by explaining the current significance of reconfigurable antennas in 5G applications.

Chapter II presents an exhaustive literature review on the history and state of the art reconfigurable antennas, 5G MIMO antenna systems.

Chapter III discusses a compact and closely spaced four-port MIMO antenna for 5G applications.

Chapter IV explains a slotted compact Four-port truncated ground structured MIMO antenna for Sub-6 3.4 GHz for 5G applications.

Chapter V designs a High-performance Quad-port defected ground-structured MIMO antenna system for 5G Sub-6 GHz RF devices.

Chapter VI introduces Design and analysis of a Novel-shaped Hexa-Band Frequency reconfigurable antenna integrated with PIN diode for S-band and 5G Sub-6 GHz applications.

Finally, **Chapter VII** presents conclusion of the thesis. The scopes for future works are also explained in this chapter.

References

[1]. Bria F.G, "4th Generation Wireless Infrastructures: Scenarios and Research Challenge", *IEEE Personal Communications*, Vol. 8, No. 1, pp. 1742-1746, **2010**.

[2]. Rappaport T. S, "Wireless Communications, Principles and Practice", *Second Edition, Pearson*, Prentice Hall, U.S.A, **2002**.

[3]. Tse D and Viswanath P, "Fundamentals of Wireless Communication", First Edition, *Cambridge University Press*, U.S.A, **2005**.

[4]. Holma H and Toskala A, "WCDMA for UMTS, Radio Access for Third Generation Mobile Communications", Third Edition, *John Wiley & Sons*, New Jersey, U.S.A, **2004.**

[5]. Vanghi V, Damnjanovic A and Vojcic B, "The CDMA 2000 System for Mobile Communications: 3G Wireless Evolution", *First Edition*, Prentice Hall, U.S.A, 2004.

[6]. Bi S and Zhang R," Wireless powered communication networks: An overview", *IEEE Wireless Communication*, Vol. 23, No. 2, pp. 10–18, Apr. **2016**.

[7]. Simsek M.D, Zhang D, Öhmann D, Matthé M and Fettweis G, "On the Flexibility and Autonomy of 5G Wireless Networks", *IEEE Access*, Vol.5, pp. 22823-22835, **2017**.

[8]. Schulz P, Matthe M, Klessig H, Simsek M, Fettweis G, Ansari J, Ashraf S.A, Almeroth B, Voigt J, Riedel I, Puschmann A, Thiel A M, Muller T and Windisch M, "Latency Critical IoT Applications in 5G: Perspective on the Design of Radio Interface and Network Architecture", *IEEE Communications Magazine*, Vol. 55, No. 2, pp. 70-78, **2017**.

[9]. Joham M, Nossek J.A and Utschick W, "Linear Transmit Processing in MIMO Communications Systems", *IEEE Transactions on Signal Processing*, Vol. 53, No. 8, Aug. **2005.**

[10]. Sharawi, Mohammad S., "Printed Multi-Band MIMO Antenna Systems and Their Performance Metrics [Wireless Corner]." *IEEE Antennas and Propagation Magazine*, Vol. 55, No. 5, pp. 218–232,**2013**; doi:10.1109/map.2013.6735522.

[11]. Syrytsin I, Vesterager H, Nørgaard S, Thomsen L, Barrio S. C and Pedersen G, "Pattern-Reconfigurable Mobile Terminal Antenna System for MIMO and Link Stabilization in LTE", 12th European Conference on Antennas and Propagation (EuCAP 2018), **2018**; doi:10.1049/cp.2018.0892.

[12]. Z. Y. Li, Z. W. Du, and K. Gong," Compact Reconfigurable Antenna Array for Adaptive MIMO Systems," *IEEE Antennas and Wireless Propagation Letters*, Vol. 8, pp. 1317-1320, **2009**; doi: 10.1109/lawp.2009.2038182.

[13]. C. Rhee," Pattern-Reconfigurable MIMO Antenna for High Isolation and Low Correlation," *IEEE Antennas and Wireless Propagation Letters*, Vol. 13, pp. 1373-1376, **2014**; doi: 10.1109/lawp.2014.2339012.

[14]. Vannithamby R and Talwar S (eds)," Towards 5g: Applications, Requirements and Candidate Technologies", John Wiley & Sons, *Mobile and wireless communications*, **2017**.

[15]. Albreem M A, Juntti M and Shahabuddin S, "Massive MIMO detection techniques: a survey", *IEEE Communications Surveys & Tutorials*, 21, 3109–3132, **2019**.

[16]. Wang X, Kong L, Kong F, Qiu F, Xia M, Arnon S and Chen G, " Millimeter wave communication: a comprehensive survey", *IEEE Communications Surveys & Tutorials*, 20, 1616–1653, **2018**.

[17]. Das A and Kolangiammal S," Performance analysis of millimeter wave communication system using 256-QAM and 512-QAM techniques", *International Conference on Communication and Signal Processing (ICCSP), IEEE*, pp. 0360–0364, **2017**.

[18]. Wang J, Zhao L, Hao Z, and Jin J, "A wideband dual-polarized omnidirectional antenna for base station/WLAN", *IEEE Transactions on Antennas and Propagation*, 66, 1–1, **2017**.

[19]. Esmail B, Majid H, Dahlan S, Zainalabidin Z, Himdi M, Dewan R, Rahim MKA and Ashyap AYI," Reconfigurable Metamaterial structure for 5G beam tilting antenna applications", *Waves in Random and Complex Media*, 1–14, **2020**.

[20]. Hussain S, Qu SW, Zhou WL, Zhang P and Yang S, "Design and fabrication of wideband dual-polarized dipole array for 5G wireless systems", *IEEE Access*, 8, 1–1, **2020**.

[21]. Li Z, Sun Y, Yang M, Wu Z and Tang P, "A broadband dual polarized magneto-electric dipole antenna for 2G/3G/LTE/WiMAX applications", *Progress In Electromagnetics Research C*, 73, 127–136, **2020**.

[22]. Yin J, Wu Q, Yu C, Wand H and Hong W, "Broadband endfire magnetoelectric dipole antenna array using SICL feeding network for 5G millimeter-wave applications", *IEEE Transactions on Antennas and Propagation*, 67, 4895–4900, **2019**.

[23]. Sun K, Yang D and Liu S, "A wideband hybrid feeding circularly polarized magneto-electric dipole antenna for 5G Wi-Fi", *Microwave and Optical Technology Letters*, 60, 1837–1842, **2019**.

[24]. Liu D, Luo H, Zhang M, Wen HL, Wang B and Wang J, "An extremely low-profile wideband MIMO antenna for 5G smart-phones", *IEEE Transactions on Antennas and Propagation*, 67, 1–1, **2019**.

[25]. Ullah H and Tahir F, "A novel snowflake fractal antenna for dual-beam applications in 28 GHz band", *IEEE Access*, 8, 1–1, **2020**.

[26]. Deng J, Li J, Zhao L, and Guo L, "A dual-band inverted-F MIMO antenna with enhanced isolation for WLAN applications", *IEEE Antennas and Wireless Propagation Letters*, 16, 1–1, **2017**.

[27]. Liu D, Zhang M, Luo H, He J, Wen HL and Wang J, "Dual-band platform-free PIFA for 5G MIMO application of mobile devices", *IEEE Transactions on Antennas and Propagation*, 66, 1–1, **2018**.

[28]. Dixit AS and Kumar S, "The enhanced gain and cost-effective antipodal vivaldi antenna for 5G communication applications", *Microwave and Optical Technology Letters*, 62, 2365–2374, **2020**.

[29]. Dixit AS and Kumar S, "A miniaturized antipodal vivaldi antenna for 5G communication applications", *7th International Conference on Signal Processing and Integrated Networks (SPIN), IEEE*, pp. 800–803, **2020**.

[30]. Tiwari N and Rao TR, "Substrate integrated waveguide based high gain planar antipodal linear tapered slot antenna with dielectric loading for 60 GHz communications", *Wireless Personal Communications*, 97, 1385–1400, **2017**.

[31]. Dixit A and Kumar S, "A survey of performance enhancement techniques of antipodal vivaldi antenna", *IEEE Access*, 8, 1–1, **2020**.

[32]. Goel T and Patnaik A, "Novel broadband antennas for future mobile communications", *IEEE Transactions on Antennas and Propagation*, 66, 1–1, **2018**.

[33]. J. Lee, J.-K. Han, and J. (Charlie) Zhang, "MIMO Technologies in 3GPP LTE and LTE Advanced", *EURASIP J. Wirel. Commun. Netw.*, Vol. 2009, No. 1, Art. No. 1, Dec. **2009**; doi: 10.1155/2009/302092.

[34]. Q. Li, "MIMO techniques in WiMAX and LTE: a feature overview", IEEE Commun. Mag., Vol. 48, No. 5, pp. 86–92, May **2010**; doi: 10.1109/ MCOM .2010. 5458368.

[35]. E. Fishler, A. Haimovich, R. Blum, D. Chizhik, L. Cimini, and R. Valenzuela, "MIMO radar: an idea whose time has come", in *Proceedings of the 2004 IEEE Radar Conference,* pp. 71–78, Apr. **2004**; doi: 10.1109/NRC.2004.1316398.

[36]. J. Li and P. Stoica, "MIMO Radar with Colocated Antennas", *IEEE Signal Process. Mag.*, Vol. 24, No. 5, pp. 106–114, Sep. **2007**; doi: 10.1109/MSP.2007.904812.

[37]. E. Fishler, A. Haimovich, R. Blum, R. Cimini, D. Chizhik, and R. Valenzuela, "Performance of MIMO radar systems: advantages of angular diversity", *Conference Record of the Thirty-Eighth Asilomar Conference on Signals, Systems and Computers*, Vol. 1, pp. 305-309, Nov. **2004**; doi: 10.1109/ACSSC. 2004. 1399142.

[38]. E. Fishler, A. Haimovich, R. S. Blum, L. J. Cimini, D. Chizhik, and R. A. Valenzuela, "Spatial Diversity in Radars-Models and Detection Performance", *IEEE Trans. Signal Process.*, Vol. 54, No. 3, pp. 823–838, Mar. **2006**; doi: 10.1109/ TSP.2005.862813.

[39]. G. Geraci, A. Garcia-Rodriguez, L. Galati Giordano, D. López-Pérez and E. Björnson, "Understanding UAV Cellular Communications: From Existing Networks to Massive MIMO", *IEEE Access*, Vol. 6, pp. 67853–67865, **2018**.

[40]. M. Badi, J. Wensowitch, D. Rajan, and J. Camp, "Experimentally Analyzing Diverse Antenna Placements and Orientations for UAV Communications", *IEEE Trans. Veh. Technol.*, Vol. 69, No. 12, pp. 14989–15004, **2020**; doi: 10.1109/ TVT. 2020.3031872.

[41]. Y. Zeng, R. Zhang, and T. J. Lim, "Wireless communications with unmanned aerial vehicles: opportunities and challenges", *IEEE Commun. Mag.*, Vol. 54, No. 5, pp. 36–42, May **2016**; doi: 10.1109/MCOM.2016.7470933.

[42]. M. S. Sharawi, "Printed Multi-Band MIMO Antenna Systems and Their Performance Metrics [Wireless Corner]", *IEEE Antennas Propag. Mag.,* Vol. 55, No. 5, pp. 218–232, Oct. **2013**; doi: 10.1109/MAP.2013.6735522.

[43]. C.-Y. Hsu, L.-T. Hwang, F.-S. Chang, S.-M. Wang, and C.-F. Liu, "Investigation of a single-plate π-shaped multiple-input–multiple-output antenna with enhanced port isolation for 5 GHz band applications", *IET Microw. Antennas Propag.*, Vol. 10, No. 5, pp. 553–560, **2016**; doi: 10.1049/iet-map.2015.0511.

[44]. K. Boyle, "Radiation patterns and correlation of closely spaced linear antennas", *IEEE Trans. Antennas Propag.*, Vol. 50, No. 8, pp. 1162–1165, **2012**; doi: 10.1109/TAP.2002.801367.

[45]. S. H. Chae, S. Oh, and S.-O. Park, "Analysis of Mutual Coupling, Correlations, and TARC in WiBro MIMO Array Antenna", *IEEE Antennas Wirel. Propag. Lett.,* Vol. 6, pp. 122–125, **2007**; doi: 10.1109/LAWP.2007.893109.

[46]. T. Taga, "Analysis for mean effective gain of mobile antennas in land mobile radio environments", *IEEE Trans. Veh. Technol.,* Vol. 39, No. 2, pp. 117–131, May **2019**; doi: 10.1109/25.54228.

[47]. M. P. Karaboikis, V. C. Papamichael, G. F. Tsachtsiris, C. F. Soras, and V. T. Makios, "Integrating Compact Printed Antennas onto Small Diversity/MIMO Terminals", *IEEE Trans. Antennas Propag.*, Vol. 56, No. 7, pp. 2067–2078, Jul. **2008**; doi: 10.1109/TAP.2008.924677.

[48]. I. Nadeem and D.-Y. Choi, "Study on Mutual Coupling Reduction Technique for MIMO Antennas", *IEEE Access*, Vol. 7, pp. 563–586, **2019**; doi: 10.1109/ ACCESS.2018.2885558.

[49]. M. Manteghi and Y. Rahmat-Samii, "Broadband characterization of the total active reflection coefficient of multiport antennas", *IEEE Antennas and Propagation Society International Symposium*, Digest, Held in conjunction with: USNC/CNC/URSI North American Radio Sci. Meeting (Cat. No.03CH37450), vol. 3, pp. 20–23, Jun. **2003**; doi: 10.1109/APS.2003.1219779.

[50]. J Youngs, G. C. Stevens and A. S. Voughan, "Trends in dielectric research: an international review from 1980-2004," *J.Phys. D: Appl. Phys*, pp 1267-1276, **2006**.

[51]. M.G.Pecht, G.R Agarwal, P.McCluskey, T.Dishongh, S. Javadpour and R.Mahajan," Electronic Packaging Materials and there properties", *CRC Press*, London, **1999**.

[52]. T.Hu,J. Juuti, H. Jantunen and T. Vilkman, "Dielectric properties of BST/Polymer composites", *J.Eur.Ceram.Soc.,* pp 3997-4001, **2007**.

[53]. D.D.L.Chung, "Materials for Electronic Packaging," Butterworth Heinemann, Washington, **1995**.

[54]. M.T. Sebastian," Dielectric materials for wireless communications", *Elsevier publishers*., UK, **2008**.

[55]. Rao Y, Qu J, Marinis T, Wong C P, "A precise numerical prediction of effective dielectric constant for polymer ceramic composites based on effective medium theory" *IEEE Trans. Compon. Packag. Tech*, pp. 680-683, **2000**.

[56]. Prakash A, Vaid J K Mansingh A, "Measurement of dielectric parameters at microwave frequencies by cavity perturbation technique", *IEEE Trans. Microwave Theory and Techniques*, Vol.27, pp. 791-795, **2015**.

[57]. HP8510C Network Analyzer operating and programming manual, Hewlett Packard, **1988**.

[58]. E. J. Zachariah, K. Vasudevan, P. A. Praveen Kumar, P. Mohanan and K. G. Nair," Design, Development and Performance Evaluation of an Anechoic Chamber for Microwave Antennas Studies", *Indian Journal of Radio and Space Physics*, Vol. 13, pp. 29-31, **2018**.

[59]. C. A. Balanis, "Antenna Theory: Analysis and Design", Second Edition, *John Wiley & Sons Inc.*, **1982.**

[60]. John D. Kraus, Antennas Mc. Graw Hill International, second edition, **1988.**

[61]. Cellular Handset Antenna Efficiency Measurement Using the Wheeler Cap, Skyworks Solutions, Inc, **2005**.

Chapter II

CHAPTER II

LITERATURE REVIEW

Today, a reconfigurable antenna is quickly emerging as a significant element of communication equipment. The antenna in contemporary communication systems should be able to operate in more than one operational mode within a single design. Reconfigurable antennas are thus created to both address the issue and provide new capabilities for adapting in future wireless applications. In this chapter, the Section 2.1 involves the Review on different reconfigurable antenna techniques in 5G is discussed in detail. The example for each technique is also presented in this section. In Section 2.2, Review of different MIMO antennas is presented which is an application of reconfigurable antenna, includes some of the different MIMO antennas with its different applications. Ansys HFSS and CST is applied vastly in the development of the antenna design in this study.

2.1 A REVIEW ON DIFFERENT RECONFIGURABLE ANTENNA TECHNIQUES

Various real application strategies have been presented to be employed in different wireless systems to identify the reconfigurable antennas, and some of them are Satellite, MIMO, Biomedical, and Cognitive radio communications, which are classified as follows:

- Changes to the electrical system
- Reconfiguration of the optical system
- Reconfiguration of mechanical systems
- Tunable Materials for Switching

[1] Qasim H. Kareem et al., presents an octagonal ring-shaped frequency tunable dual band reconfigurable antenna for the application of Wireless Communications. The antenna has four octagonal ring-shaped radiators present in the ground plane. To provide frequency reconfigurability, a varactor diode is given in the center of each of the radiators to control the slot capacitive reactance. The antennas are orthogonally positioned and the rectangular defective ground gaps are used for the purpose of isolation. Dual band operation is achieved and it is applicable for future wireless communication systems.

[2] Amjad Iqbal et al., describes a printed and a compact sized antenna frequency as well as pattern reconfigurable antenna. The antenna consists of three switches. Operating bands of the antenna is controlled by switch 1. Other two switches control the beam switching of the antenna. The given antenna operates at 3.1 GHz and 6.8 GHz when the switch 1 is in ON condition and the antenna operates only on 3.1 GHz when the switch 1 is in OFF state. By adjusting ON and OFF states of the other two switches, the desired beam from the antenna can be obtained. By giving different combination of both the switches, unique beams can be obtained.

[3] Mohammad M. Fakharian et al., presents a monopole fed reconfigurable loop antenna using RF PIN based switches and shorted parasitic elements. This antenna can be used for GPS applications. It can reconfigure multiple polarizations. The metallic patches are used for parasitic elements which provides a capability for reconfiguration to the antenna and acts as a driven monopole fed looping. The patches can connect or disconnect by using switching and the antenna can simultaneously change both the planes and also switches among all the types of polarizations in different modes.

[4] Mustafa Murat Bilgic et al., demonstrated a polarization reconfigurable antenna for WLAN and Wireless sensor applications. It is having a degenerating dual feed patch used for the circular polarization and L-shape can either connect or disconnect to the main area of the patch and RF switches can be placed around the patch. The patch will radiate in linearly polarizable either vertically or horizontally when the L-shape is connected and also it depends on the feed that is being used. When the switches are not biased, then the antenna is linearly polarized.

[5] Sunghwan Kim et al., describes a simple structured frequency switchable antenna having eight frequency choices. Totally three PIN diodes are used in their chosen places for the antenna resonant bands shifting. Depending upon the ON and OFF states of the PIN diode, the antenna operates at four different modes. In each mode, it covers two frequencies in a unique way. The given antenna may be combined also in heterogeneous RF front ends.

[6] Phalguni Mathur et al., shows a mechanically frequency reconfiguration techniques for WSN applications. The frequency reconfiguration can be achieved by changing the dielectric properties of the antenna mechanically and obtains a tunable frequency range from 2.4 GHz to 2.7 GHz. The gain of the antenna remains flat during the reconfigured frequency range with the radiation pattern stabled. The fluid level sensor will operate based on the resonant frequency variation with the rise and fall in liquid. Thus, the antenna can able to sense and can transmit the information concurrently.

[7] Taehee Jang et al., presents an optically transparent Zeroth-order resonant mechanically reconfigurable antenna using metallic micromesh. Reduction in size of the antenna is achieved by using the given resonant property and the metallic patches are replaced with the micromesh. Upon stretching, the resonant frequency is linearly configurable from 2.94 GHz to 2.46 GHz. The structure undergoes a mechanical bend such as without breaking, it stretches, fold or twist.

[8] I F da Costa et al., offers a millimeter-wave optically patch antenna for usage in future broadband cellular communication networks such as 5G application. Silicon switches are used to monitor the reconfiguration. Additionally, it modifies the antenna design's emission pattern and frequency range. In 5G networks, the system's ability to switch between the 28 GHz and 38 GHz frequency bands is highly helpful. The poor constellation for the OFF state improves as the switch is turned to the ON position, and this result is very encouraging for untapped frequency bands outside of commercial systems.

[9] A M Mabrouk et al., describes a 5.1 GHz antenna with a multiple radiating pattern. A copper ground layer supports the antenna's four centre-fed copper arms, which are distributed across a substrate. Then, by placing transformer oil within a container between the substrate and the ground, this antenna is altered into a frequency-reconfigurable antenna. The revised antenna may be maintained at multiple frequency ranges between 5.2 GHz and 6.5 GHz by varying the height of the oil in the container. Four PIN diodes are inserted between the four arms of the redesigned antenna and the feeding point to provide beam reconfigurability. Finally, a number of reconfigurable antennas with changing gain and HPBW values are made in a different orientation at various operating frequencies.

[10] Sarawuth Chaimool et al., presents a meta-surface antenna that is changeable. It is composed of equally spaced double-split square rings that are loaded with PIN diodes, which serve as both a wave controller and a wave reflector. When the PIN diodes are biased in the forward direction, an epsilon-negative meta-surface is realised, showing all incident waves to have proper polarisation; when the diodes are biased in the reverse direction, a mu-near-zero meta-surface is realised, manipulating wave propagation at the same operating frequency. For excitation, a dipole radiator loaded with the same kind of PIN diode is used. Applications involving beam switching can exploit this.

[11] Ming-Chun Tang et al., describes a modular, frequency-reconfigurable filtenna with sharp out-of-band rejection, which can be used in both its wideband and reliably adjustable narrowband forms. It is designed for applications involving cognitive radio. The wideband state, which operates between 2.3 and 4.9 GHz, is used for sensing. The WiMAX band and the C-band are included in the narrowband states, which are intended to cover communications between 3.0 and 4.3 GHz. A PIN diode is utilised to switch between these wideband and narrowband operational states. The operational frequencies are continuously switched between the narrowband states using two varactor diodes. A monopole in the form of a funnel and a repositionable filter makes up the filtenna.

[12] Mahmoud Al Ahmad et al., displays a single variable-frequency antenna with a varactor. Four metallized patches in a symmetrical pattern, connected and joined by slots and metallic bars, make up the antenna. The antenna arrangement is excited by a single feed probe that is affixed to one of the patches. A shorting pin near the feed point is used to reduce the resonance frequency of the antenna in order to obtain a tiny dimension. The centre of the patch is equipped with a varactor diode to provide frequency reconfigurability. A tuning range is made possible by variations in the DC reverse bias voltage of the varactor's resonant frequency. The design is straightforward compared to the design features listed in earlier research, and it executes frequency reconfiguration using a single varactor diode and a low bias voltage consumption.

[13] Hesham A. Mohamed et al., introduces a unique flower-shaped microstrip line feed reconfigurable band-notched UWB monopole and built utilising a single varactor diode. Using varied capacitance values, different notch frequencies can be generated.

The effect of shifting the location of the varactor is also investigated. To get UWB properties, the flower shape is first tuned. The microstrip line is then slotted to accommodate a single varactor diode. This simple setup, which encompasses the majority of narrow band coexisting systems, yields a wide range of notch frequencies. By raising the capacitance value, the notch frequency can be reduced. With the same antenna arrangement and varactor location, the notch frequency includes the WLAN band when C = 0.8 pF and the WiMAX band when C = 0.7 pF. Two prototypes of the proposed antenna are built using two distinct single capacitor elements with capacitances of 0.6 pF and 1.5 pF, and their reflection characteristics are examined and compared to simulation results. In both computed and measured antenna configurations, notch frequencies of 6.1 GHz and 4.3 GHz are obtained. The given antenna is good for modern UWB systems.

[14] Preet Kaur et al., recommends a novel hybrid reconfigurable fractal antenna that combines the advantages of wideband and multiband applications. The fractal structure's identity causes the supplied antenna to be multiband in each reconfigurable mode. The antenna has only three resonant frequencies if a non-reconfigurable fractal structure is employed, in which the centre patch is fully attached to all small patches; however, employing the reconfiguration concept, this antenna can empathise at about ten different resonant frequencies throughout all modes. When using the reconfiguration method, power interruption issues can also be avoided by switching between different modes. The suggested antenna has uses in wireless industrial, medical imaging, microwave imaging, vehicle radar, and satellite communication.

[15] Phalguni Mathur et al., describes about novel mechanical reconfigurable antenna for Internet of things applications. The rectangular microstrip patch antenna has four end to end cavities placed upon a dielectric slab. The cavities are seeming as binary bits. Each bit depends upon the materials filled inside it. There are total of 16 combinations each of it having unique characteristics. FR4 material is used as substrate and the same material is used to fill the cavities in the first case. In the second case, Rogers RT/duroid 6010/6010LM (tm) material is filled. The variation range of resonant cavity can be high if the dielectric constant material as filler is higher. This can be used as wireless sensor network-based applications.

[16] Sonia Sharma et al., proposes a novel frequency reconfigurable U-slot antenna with T-shaped feed line used for SDR communication systems. It operates at three different frequencies depending upon the switching states of PIN diodes. Another improved antenna is also designed to strengthen the proposed antenna. By comparing both the antennas, the second antenna utilizes four slots to increase the number of reconfiguration states. This antenna operates on five different frequency bands. The radiation pattern is stable at all the operating frequencies.

[17] Ting-Yan Liu et al., demonstrates a dual band frequency reconfigurable antenna. The radiating patch of the provided antenna is equipped with an annular slot and is shorted to the ground. There are three feed ports in total. One of the three ports is used to generate the monopolar patch mode resonant at a higher frequency, while the other two ports are used to excite the slot mode resonant at a lower frequency. Dual orthogonal linear polarization is present in both of the first two ports, and vertical polarization is present in the third port. To reposition the annular slot, there are four symmetrically positioned varactors. When the capacitance of the varactors is changed, the results of the simulation show the slot mode resonant frequency. When an antenna uses the monopolar mode, its impedance bandwidth can cover the frequencies in every capacitance value.

[18] Mubashir Ul Hassan et al., presents a novel flexible antenna with frequency reconfigurability for quintuple applications. The Rogers RT/duroid 5880 is used as the substrate for the suggested antenna. On the antenna's surface, two switches have been integrated to change the frequency and current distribution under various switching conditions. The provided antenna has a VSWR under two, a strong radiation pattern, and positive gain over the specified frequency bands. It can be used for a variety of applications, including WLAN and X-band.

[19] Guiping Jin et al., offers a selectively fed frequency reconfigurable antenna for 5G and WLAN applications operating at sub-6 GHz. A 2.5mm thick air gap separates the two substrates. Four PIN diode switches with two separate frequency bands can be produced on the two substrates. The feed line length and the excitation of the radiating patch can both be adjusted at the same time. The radiation patterns and gains for both states are comparable according to both the simulated and measured data.

[20] Geriki Polaiah et al., showcases a slot line open loop resonator-based frequency-configurable antenna. This antenna has two ports; port 1 receives the control signal and transforms it to a DC signal using a rectifier; the rectified signal is then used to shift the second port's frequency bands. The antenna has been switched from lower to higher frequency bands when the PIN diode is ON, according to measurement results. Without an external DC bias voltage, the antenna in its previous configuration might be acceptable for transmitter reconfigurable antenna.

[21] Abu Hena Murshed et al., discusses a frequency-reconfigurable honey bee antenna that operates in dual and triple bands in eight different modes and is fed by a microstrip line. Three PIN diodes are inserted over the honey bee arms, controlling the effective current distribution in eight different ways. This might be more appropriate for use with cognitive radio. This can also be used for a variety of wireless applications, including WLAN, Wi-Fi, and WiMAX. The antenna is printed with a planar geometry on a flexible FR4 substrate. For all resonance frequencies with a positive gain, the provided antenna displays VSWR less than two.

[22] Subash et al., introduces a compact four band frequency reconfigurable antenna for satellite and 5G applications. The antenna consists of two rectangular slots in the radiating patch which resonates independently at both the single and quad bands. At the middle of the rectangular slot, a metallic switch is placed which helps in achieving the frequency reconfigurability and hence when the switch is on ON condition, the antenna resonates at single band and when it gets OFF, it resonates at Quad band. The antenna has good impedance matching and the radiation pattern and the gain are much more stable.

[23] Malathi Kanagasabai et al., describes a WBAN application-specific frequency reconfigurable antenna. It makes use of the switching capabilities of PIN diodes to achieve frequency reconfigurability. The suggested antenna is made up of three distinct frequency bands, a modified loop-shaped resonator, a U-shaped stub, and an F-shaped strip that is mirror imaged. GPS, ISM, and wireless fidelity applications all employ this antenna. For off-body communication, the resultant directed radiation pattern is ideal. Applications for wearable technology can also exploit this.

[24] Haris Dildar et al., presents a monopole frequency-configurable antenna. It is constructed on a FR4 substrate, and four PIN diodes are put between the radiating patches to enable switching between operational modes. It uses single bands in modes 1 and 2, dual bands in modes 3 and 4, and triple bands in mode 5 to operate across nine separate bands. The developed antenna has good efficiency and VSWR is less than 2 for all operating frequencies. It can be utilised for 5G and Internet of Things applications due to its reasonable size and ability to support numerous wireless standards.

[25] Rahul Kumar Verma et al., proposes a frequency reconfigurable antenna which is compact and coplanar waveguide fed integrated with PIN diode for 5G and UWB applications. The design is made up of an FR4 substrate in which the patch length is altered due to diode, acts as a switch with proper biasing. For validating the characteristics of the antenna, all the parameters are verified. When the diode gets ON, the operating frequency ranges from 3.1 to 3.6 GHz and 6 to 8.8 GHz and if it is in OFF condition, the frequency ranges from 3.8 to 5.05 GHz and 7.55 to 9 GHz respectively. The results are much useful for 5G applications.

[26] Poonam Thanki et al., describes a fork-shaped antenna that can change frequency and pattern. The suggested antenna's frequency and pattern are reconfigured using four PIN diodes. The suggested antenna operates at multiple frequencies, and has a pattern tilt in various PIN diode switching modes. A high-frequency structural simulator is used to simulate the proposed antenna. To reconfigure pattern and frequency in HFSS, lumped RLC elements are used. To test the simulated results, PIN diodes are included in a produced prototype for switching. In both simulated and measured findings, the suggested antenna exhibits satisfactory matching.

[27] M. Faizal Ismail et al., features a pattern reconfigurable antenna that allows for various beam steering configurations to handle the issue of overcrowded signal situations while maintaining efficiency and operating frequency and enhancing signal and data security. For beam steering configurations, a pattern-reconfigurable antenna with an EBG structure is described. The PIN diode on the radiating patches was advised in the majority of earlier research, which decreased their performance. Since the RF switches are not incorporated into the radiating patches, the structure of this design is efficient.

Designing and analysing a dual-band slotted pattern reconfigurable antenna is done using the EBG structure. To the right and left of the antenna's radiating edge are where the EBG patches are located. The EBG is designed so that its surface band gap wave covers the antenna's radiation. The given antenna is accomplished of having beam shifting at three dissimilar angles.

[28] Jerzy Kowalewski et al., describes a 2.6 GHz compact antenna system with pattern reconfigurability is designed and evaluated. Top loading lowers the antenna's height, which is based on the idea of an electrically steerable parasitic array radiator. The antenna can generate 10 reversible patterns and has a maximum gain of 7.4 dBi. Additionally, it is planned to use these antennas as part of a multiple antenna system. The radiation patterns produced by this MIMO system are ideal for vehicular urban settings, according to prior study. S-parameter measurements on a built prototype yield results that are consistent with the simulation. Gain of up to 8 dBi is achieved, and 3D explanations of radiation patterns closely match simulation.

[29] Jinzhi Zhou et al., draws attention to a dual band, pattern-reconfigurable antenna. A 2 mm air layer and two substrates are used to construct the antenna. Two radiators are printed on the top surfaces of the two substrates. The top surface of the upper substrate is used to mount the first radiator, a circular patch. No metal layer is etched on the bottom surface of the top substrate, and eight rectangular slots are additionally placed to introduce a directional radiation pattern with reconfigurable features in the low band by varying the current distribution. In order to create a poor coupling effect with the circular patch, the second radiator, which is printed on the top surface of the bottom substrate, comprises of a crossing branch and four arc-shaped branches. Towards the base of the lower substrate, a round ground plane and four symmetrical rectangular slots are engraved to provide an extra resonating point in the high band with the pattern reconfiguration characteristic. In the rectangular slots, 12 PIN diodes are inserted to test the accuracy of the dual-band and pattern reconfigurable features.

[30] Kushmanda Saurav et al., suggested a dual band design reconfigurable antenna. Two metal strips of varying lengths serve as parasitic segments on both sides of the driven dipole in the generated portion of the antenna, which is a dual-band complementary split ring resonators (CSRRs) loaded dipole. To control the electrical length of parasitic components, PIN diodes are employed. In addition to the reference dipole mode, the CSRRs concentration also provide a restricted overs mode, and the parasitic elements and PIN diodes can change the dual-band driven element's omnidirectional radiation into nine different radiation pattern configurations in both working bands. WLAN applications can make use of the included antenna.

[31] Adnan Ghaffar et al., features a flexible antenna with adjustable frequency and pattern for several applications. The frequency and pattern of a triangular monopole antenna with a semi-circular stub can be changed by connecting and disconnecting two inverted L-shaped stubs with diodes. Any one of the stubs can be used to connect to the radiator, which creates a phase difference at both ends and changes the electromagnetic radiation's direction to enable pattern reconfigurability. Additionally, due to the reactive load that the stubs added, the antenna's effective length changed, enabling frequency reconfigurability. By switching the statuses of both diodes to OFF and ON, respectively, the recommended antenna's frequency can be changed from 2.1 GHz to 1.8 GHz. To ensure that the antenna prototype's overall performance is correct. Furthermore, the proposed design was validated by comparing it to preceding artworks in terms of total size, reconfigurability type, adaptability, radio frequency (RF) switch type employed for reconfigurability, and frequency bandwidth. The designed antenna has a reduced dimension with a high bandwidth coverage, as well as a discrete RF switch type with flexibility and frequency and pattern reconfigurability.

[32] Kranti Kumar Katare et al., proposes an increased gain pattern reconfigurable antenna system. It is made up of one controlled monopole antenna in the centre, encircled by six parasitic elements. Each parasitic element is formed of a metallic strip that is loaded with a switch and is protected by a dielectric layer. For the OFF and ON states of the switch, parasitic elements will react similarly to translucent and opaque surfaces to a vertically polarised incident wave. As a result, the complete assembly can be reconfigured, and beam steering is possible. The designed scheme has a high and consistent gain of 8 dBi, as well as beam steering across the whole azimuth plane. CST Microwave Studio software was used to perform the simulation for the ISM band.

[33] Ke Li et al., proposes a special pattern-reconfigurable antenna for body area network communication based on characteristic method theory. The antenna can switch between omnidirectional and broadside emission patterns, which enables it to simultaneously support on-body and off-body communication lines. Four copper posts are inserted between the radiator and the ground to create the on-body mode, and an array of 4x2 rectangular plates with the off-body communication system at BAN frequency band is used as the main radiator. Selective activation of the two modes is achieved through a straightforward feeding network. The results from simulation and measurement are very similar. Further research is done on the specific absorption rate (SAR). In addition, the link budget of the on-body link is looked at for application. The proposed antenna is a good choice for BAN communications due to its steady impedance and radiation properties, low SAR values, and reliable communication ability.

[34] Saeed A Haydhah et al., describes a compact pattern reconfigurable antenna with four radiation patterns is demonstrated for sub-GHz IoT applications. The antenna's Mode 1 features three different patterns, while Mode 2 has omnidirectional electric and magnetic patterns. The antenna's resonance frequency is 868 MHz, and its overlapped -6 dB impedance bandwidths for modes 1 and 2 are each 22 MHz and 20 MHz. The radiating elements consist of two meandered slots and one meandered monopole. Four different pattern configurations are attained. The other two designs are omnidirectional, whereas two of the designs have a 5 dB directivity. In order to accomplish pattern reconfigurability, four PIN diodes are used. By removing the FR-4 inside the slots, the radiation effectiveness of the slots was increased by 2.25 dB.

[35] Abdellah Attalhaoul et al., analyses linear polarisation reconfigurable patch antennas for 5G applications. The first category is a U-shaped antenna that can flip between two symmetric polarizations. The second type of antenna has been studied and can flip between two orthogonal polarizations. Both categories of antennas can generate polarisation diversity while maintaining other performance characteristics such as radiation pattern and frequency. These antennas are necessary for wireless communication systems because they prevent fading and boost channel capacity. Two diodes are inserted into a microstrip feed to achieve linear polarisation reconfigurability. The resonance frequency of antennas was determined to be between 3.4-3.6 GHz in both stages of operation. This antenna is a good candidate for 5G communication.

[36] Chang Xu et al., proposes a wideband, high-gain antenna based on liquid metal tuning that is polarization-adjustable. A center-feed circular patch antenna with a C-shaped slot serves as the antenna's driving element. A parasitic circular disc and 13 circular directors are used to broaden the bandwidth and enhance the gain that is really realised. The polarisation direction can be altered by adjusting the liquid metal alloy in the tube. The relationship between the dimensions and the potential gain has been investigated. From 4.3 to 5.3 GHz, the antenna's impedance bandwidth is -10 dB. The modelling and measurement results show good agreement.

[37] Yasir I A Al-Yasir et al., provides a unified, high axial ratio, and frequency responsive circular polarisation reconfigurable antenna for 5G applications. In this setup, the reconfiguration of right-hand circular polarisation and left-hand circular polarisation is controlled by two PIN diode switches. By removing two ends from the radiating patch, a rectangular C-slot patch antenna with truncated end approaches is employed to provide reconfigurability. For both steps of reconfiguration, the suggested antenna was simulated with 3.35-3.77 GHz and 3.4-3.72 GHz bands; both are suitable for 5G applications.

[38] Zhilong Li et al., designs for an electrically down-tilt dual-polarized antenna for 5G communication applications that incorporates a reconfigurable reflecting meta-surface and a partly reflective surface. This can change the major lobe orientation of BSAA by adjusting the varactor diodes that are put into the reflective elements. Furthermore, the PRS across the array is used to build a Fabry-Perot cavity with a reflecting meta surface and ground plane. A dual-polarized BSAA running from 3.4 GHz to 3.6 GHz is designed and manufactured using this design technique. The simulated and measured findings suggest that PRS increases gain by roughly 2.56 dB while decreasing side lobe level by less than 20 dB. The antenna's main lobe can be precisely adjusted within 5° for beam down-tilt.

[39] Emad Al Abbas et al., works to create a 5G 4-element antenna array system that can switch between right- and left-handed circular polarisation. By switching two PIN-diode pairs in the feeding circuit, which consists of a branch line coupler and two feeding arms, the CP-reconfigurability is achieved. The proposed method permits the use of faulty PIN diodes at millimeter-wave frequencies while reducing the number of tuning components. Axial ratio bandwidth for the prototype is 3 dB from 27.2 to 28.35 GHz. A maximum measurement of around 6 dBic and a radiation efficiency of about 51% over the bandwidth are attained by both CP states.

[40] Naser Ojaroudi Parchin et al., shows a novel circular polarisation method. It was shown a reconfigurable antenna for 5G wireless communications. The antenna was compact, had a great axial ratio, and had a good frequency response. It also had a semi-circular slit. Two PIN diode switches were used to reconfigure the circuit for both the right- and left-hand circular polarization. Reconfigurable orthogonal polarizations were achieved by varying the states of the two PIN diode switches, but the reflection coefficient was kept, which is a key benefit of this system. In both modes of operation, it had a 3.4 GHz resonance frequency, an excellent axial ratio of less than 1.8 dB, and a decent gain of 4.8 dBic.

[41] Swetha Ravikanti et al., presents a circular polarised antenna for use in 5G WLAN. The apparatus consists of a circular patch and a coaxial probe feed radiator. At the three edges of a 90° circular route, equal semi-circular slits of radius r1 are used to provide circular polarisation (CP), while the remaining semi-circular slit has a variable radius r2. A circular patch antenna's fourth circumferential slit's radius is changed to boost the impedance bandwidth and axial ratio. The measured gain is 6.6 dBi at the resonance frequency of 5.37 GHz, the measured 3-dB axial ratio bandwidth is 1.1% (5.4-5.4 GHz), and the measured 10 dB impedance bandwidth is 3.9% (5.3-5.5 GHz). The researchers also offer two-dimensional radiation charts of the surface current distribution in the azimuth and elevation planes of the CP patch.

[42] Xue Ren et al., presents a novel selectively dual-polarized magnetic dipole antenna with a distinctive pattern, platform independence, large bandwidth, and robust structure for spaceborne applications. In the proposed antenna, a specially constructed rectangular cavity supports a dual-polarized magnetic dipole. The magnetic dipole is constructed from two parallel-plate waveguides that are perpendicular to one another and have stepped edges to increase bandwidth and give structural stability, which is crucial for spaceborne antennas. The conductor that powers the dual-polarized magnetic dipole antenna is supplied by two sets of capacity coupling feeding probes. The radiation pattern in both planes is shaped by four metal plates in each of the four corners of the specially constructed rectangular chamber. Furthermore, because of the cavity, the size of the conductive surface on which the antenna is placed has little effect on the electrical functionality of the antenna, particularly the radiation pattern. A prototype is made and tested. The measured results show that the antenna has a fractional bandwidth of 18.5% and a reflection coefficient of -10 dB. The realized gain is around 8.5 dBi.

[43] He Huang et al., offers an exceedingly small, single-ended, dual-polarized patch antenna. It consists of three substrates, each of which is in charge of feeding, transmitting energy, and radiating. A patch is printed on the top layer of the upper substrate. The patch has four triangular slots carved into it to carry current in a slanted manner and minimise hindrance to nearby structures. Then, to increase structural symmetry and connect the energy to the patch, two linked lines are put crosswise on the top surface of the medium substrate. The linked lines are connected to the transition lines on the bottom side of the lowermost substrate using plated through holes. In order to verify, the antenna element and its 1x2 sub-array are treated and tested. The results indicate that the suggested antenna has a lot of promise for 5G new radio (NR) applications.

[44] Khaled Yahya Alqurashi et al., demonstrates how large areas of metallization can be connected and detached while still obtaining radiation efficiency that is not possible with conventional switches. The antenna design can change its operational bandwidth between ultra-wideband and narrowband by joining and separating the ground plane for the feed point from that of the radiator. Typical semiconductor switches could be used for this. On the other hand, these switches have terminals that resemble points. There are electrical contact areas between the switches as a result. Back radiation is produced in large quantities by surface currents flowing across these areas. Liquid metal slugs are used to completely fill the gaps. This offers a pattern equal to a normal microstrip patch antenna while significantly reducing back radiation, improving boresight gain, and both. The realised gain and overall efficiency are both enhanced by 2 dBi and 24%, accordingly. The antenna could be used in wireless systems that use Cognitive Radio and Spectrum Aggregation.

[45] Mohamed Bikrat et al., presents the bandwidth of a reconfigurable planar antenna design functioning in 5G and ultra-wide band (UWB) applications. A two PIN diode F-shaped slot and two thin vertical slots are used to obtain switching functionality. With an impedance bandwidth of 4.41 GHz, a defective ground structure is used to improve a frequency range. Furthermore, the current flow within the patch's radiation is regulated by a pair of PIN diodes located in the patch's centre in the F-shape slot. The bandwidth re-configurability is achieved by altering the operating states of the PIN diodes in a retractable patch antenna design. As a function, a reversible Circular Polarization (CP) is achieved. Furthermore, at 3 dB Axial Ratio (AR) bandwidths, the proposed antenna achieved an efficiency of 88.55% in frequencies.

[46] T. Anilkumar et al., provides a thin frequency-configurable antenna for use in vehicle communications. The patch antenna system is made up of a liquid crystal polymer (LCP) substrate, a coplanar waveguide fed circular ring system, and orthogonally oriented parasitic circular components. By adjusting the PIN diode switching components at 2.4 GHz, 5.9 GHz, 9.5 GHz, and 12.5 GHz in the antenna design, the effective bandwidth of the antenna can be altered. This work also evaluated the far-field properties of the radome enclosed antenna after mounting the antenna on the roof of the car, as well as the impact of radome substrate permittivity and radome housing on the performance of the suggested reconfigurable antenna at various modulation states. Applications involving real-time can utilize this.

[47] Ulomi George et al., proposes a patch antenna with adjustable frequency and polarisation. The antenna measures 40x40x2.5 mm^3 and has a simplistic design. While switchable truncated edges of the patch antenna are utilised for linear-to-circular polarisation reconfiguration, PIN diodes are used for frequency reconfiguration. The suggested antenna design is intended for mobile satellite communication systems, wireless local area networks, global microwave access, and the L1 band of the global navigation satellite systems.

[48] Jun Hu et al., runs a patch antenna with a slight gain fluctuation that is frequency and polarisation changeable. Eight distinct operating frequency bands may be produced by correctly shorting Pins, and three polarisation states may be attained successively by switching disruptive segments. A steady gain effectiveness is achieved over a large reconfigurable frequency range as well as all reconfigurable polarisation states by exploiting the loss-character of p-i-n diodes when shorting pins are suitably set ON/OFF for different states. The creation, construction, and testing of a prototype. The results from the simulation and the measurement coincide. The suggested antenna has a small gain variation of less than 1.8 dB across all working states, a simple reconfigurable network, a wide switching frequency range, and a respectable element gain, according to the testing results.

[49] Suresh Kumar M et al., suggests a polarisation and frequency-configurable circular microstrip antenna. A wideband coupled branch line coupler with switching feed configuration is used to complete each polarising term in the antenna. This article proposes using a cascaded BLC feed network to control adjustable over a wideband of 1.2 to 2.9 GHz. The impedance bandwidth of the circular patch antenna can be changed between 2.0 and 2.9 GHz. It is done by varying the varactor diode's capacitance values between 4.15 pF (0V) and 0.94 pF (6V). The impedance bandwidth of the BLC feed network must be matched simultaneously with the impedance bandwidth of the patch antenna in order to operate at peak efficiency. A good level of agreement is seen from the measured results.

[50] Ghanshyam Singh et al., achieves a novel end-fire pattern programmable, lightweight, simple, and frequency antenna. The proposed antenna has a closed ring resonator, an electric-inductive-capacitive resonator, and four parasitic elements. Five PIN diodes are implanted on the substrate's top surface to give the antenna its reconfigurability. First diode is put between the ELC and CRR resonators for frequency reconfigurability. The remaining four diodes are inserted between the ground plane and four parasitic elements to modify the electrical length of the ground plane, resulting in pattern variety. The primary omnidirectional beam is guided to Bi- and Uni-directional end-fire radiation at different frequencies by the ground plane and parasitic components. The proposed multiband operation states for the provided antenna. It can also be used in back-to-back repeater systems.

[51] MLSNS Lakshmi et al., features a polyimide substrate material that is used to design an uneven ground structural hybrid reconfigurable antenna. In the radiating layout of the antenna, there are a few F-shaped slots and a U-shaped slot. The antenna's frequency range at its three resonance frequencies is (2.3-3.8 GHz), (4.5-5.1 GHz), and (6.3-6.8 GHz). Measurement analysis with relation to geometrical attributes is carried out to optimise the antenna performance requirements. The asymmetric ground with slots on the radiating structure allows frequency and pattern reconfigurability through the location of diodes and tunability by switching conditions. The suggested antenna offers S and C band frequency and pattern reconfigurability for LTE, Wi-Fi, WLAN, and fixed satellite applications. Antenna efficiency of more than 74% and a peak gain of 4 dB are both attained in the antenna's fundamental resonant area. Four state frequency reconfigurability and two state pattern reconfigurability are achieved with beam steering in the azimuthal plane. With strong agreement found between simulation results produced and measurement results obtained from the combinational analyzer in the anechoic chamber, the adaptive response of the antenna is also examined at various bending angles in the X and Y directions.

[52] Anuradha A Palsokar et al., achieves the design of a frequency and pattern reconfigurable rectangular patch antenna utilising an only one PIN diode switch. The usage of a single PIN diode simplifies the complication of the diode's biassing network. The suggested antenna operates in two modes and has resonance frequencies of 2.47 GHz, 3.8 GHz, and 5.36 GHz, as well as the ability to adjust the radiation pattern. Biasing lines required for diode biassing are located distant from the radiating structure in this design, and the correlation-based technique is employed to increase antenna feed for impedance matching.

[53] Yixin Li et al proposes the standard reconfigurable antennas, which only use PIN diodes as switches, a unique approach for obtaining reactance tuning in which the impedance features of the PIN diode in different operating modes are used to realise component information exchange. First, mathematical analysis is used to describe the mechanism of element sharing. The unique way of implementing element sharing based on PIN diode impedance characteristics is then explained. Using a PIN diode in this manner, reactance adjustment is simple. Finally, to demonstrate the viability of the suggested technology, a 360° beam switchable dipole antenna array is developed.

[54] Son Xuat Ta et al., introduces four parasitic patches that are arranged to create low-profile broadband antennas between planar radiators and an electric conductor ground plane. Linearly and circularly polarised radiations are produced using a standard dipole and a crossing dipole, respectively. The radiator and parasitic patches are used on thin substrates to reduce expenses. In terms of profile minimization, resonant frequency reduction, and bandwidth enlargement, parasitic patches improve antenna performance. These improvements are described and shown via computing.

[55] Youcef Braham Chaouche et al., showcases a reconfigurable coplanar waveguide antenna with fractal geometry in the form of a crescent. The frequency reconfigurable approach is implemented using RF-PIN diodes, resistors, and inductors. Reconfigurable switching over eight frequency bands, ranging from 1.4 to 6.1 GHz, was successfully made possible using the suggested technique. Good results have been reached in terms of stable and omnidirectional radiation patterns. The antenna system's achieved gains in these frequency bands range from 0.52 to 5.67 dBi. The constructed antenna has a multiband structure and a small dimension, which are advantages over previously reported structures. Future wireless application systems are compatible with the antenna that is being described.

[56] Harish Chandra Mohanta et al., explains the functions and design process for reconfigurable antennas. It discusses the activation procedures for reconfigurable antennas as well as their design and operational optimization. The most recent developments in reconfigurable metamaterial engineering are addressed, along with current patterns and potential applications in reconfigurable antennas. Finally, the use of reconfigurable antennas in cognitive radio, MIMO systems, satellite communications, and medical equipment is examined.

[57] Abdellatif Bouyedda et al., suggests that because of its lower prices and global coverage, UHF satellite connectivity for Internet of Things technology is quickly gaining favour in monitoring applications. While low power wide area network solutions are now available, they include white zones. In this research, the authors suggest a reconfigurable UHF antenna for use with the Low Earth Orbit (LEO) Kineis constellation for LoRaWAN and UHF satellite communication. The antenna's resonant frequencies are controlled by a PIN diode and are built with a meandering line structure filled with lumped components.

[58] Jeong Keun Ji et al., discusses a dual-band pattern reconfigurable antenna for wireless MIMO (multiple-input multiple-output) applications at 2.4 and 5.8 GHz. The suggested antenna consists of four combinations of loaded PIN diodes that are both active and parasitic. By changing PIN diodes, the parasitic element functions as a reflector, allowing the antenna's radiation patterns to be modified. The antenna features three techniques and nine radiation beam patterns for the 5.8 GHz band. Peak gains for all the investigated beam patterns ranged from 5.6 to 10.4 dBi. In the 2.4 GHz band, omnidirectional beam patterns are created with an observed peak gain of about 4.5 dBi.

[59] Muhammad Ikram et al., demonstrates a dual-purpose slot antenna in the mm-wave and microwave bands. The structure's ground plane has a space printed on it as part of the design. To provide frequency configuration from 2.0 to 2.7 GHz, a short-circuited varactor diode with a peak realised gain of 4.5 dBi is used. With a bandwidth of 23-29 GHz and a maximum attainable gain of 12.5 dBi, the slot functions as a connected slot antenna array in the mm-wave range. To increase functionality, two slots are placed orthogonally for MIMO application. Future mobile portable devices will find the presented antenna system appealing since it combines 4G and 5G operation into a single small slot and has a compact, basic, and simple hierarchy.

[60] Qasim Awais et al., introduces and implements a dual-band coplanar waveguide fed printed antenna employing a FR4 substrate with rectangular shape design parts. The impedance bandwidth and gain of the antenna are enhanced via a novel rounded corners method. The intended antenna design includes two ultrawide bands, including the 2.4 GHz Bluetooth/Wi-Fi frequency, the majority of 3G and 4G bands, and the 3.4-3.6 GHz range that is anticipated to be used for 5G in the future. It is perfect for upcoming 5G Internet of Things (IoT) portable applications because it has a small antenna.

[61] Arjuna Muduli et al., shows how a fabric substrate was used to construct a reconfigurable wearable antenna. The developed antenna might work with a frequency change to operate in the 5G mid band. The suggested antenna can be used for Wi-Fi,

WLAN, satellite, 5G mobile, and cognitive radio communication and can be carried as a part of a uniform or any other type of apparel. By combining all of the above-mentioned reconfiguration elements into a single antenna and performing a parametric analysis, the antenna shape was created. This may enable successful results when using fabric like polyester as the substrate.

[62] Mingyu Sun et al., introduces a special multiband frequency-variable antenna. The designed antenna can operate in L-band at a single frequency and in S-band at two frequencies by electronic manipulation. By connecting a ground plane T-shaped slot and a ring slot, respectively, radial ring and sawtooth patches function as L/S-band radiators and are actuated by a wideband reconfigurable orthogonal network. By changing the bias of the PIN diodes that are mounted on the feed network, the operating frequency can be changed between the navigation and satellite link receiving states. The creation of a circularly polarised antenna with a highly adjustable frequency ratio creates opportunities for multipurpose CP antenna aperture sharing.

[63] Muhammad Yousaf et al., creates a low-profile antenna system for ISM band-based biomedical applications. By using a shorting technique, a circular cut-shaped radiator and polygon-shaped open-ended slots in the ground plane are provided. The proposed antenna, a novel-shaped single wide band antenna with -12.98 dBi gains at 2.45 GHz, is mounted on the homogeneous skin phantom to evaluate the outcomes. The results show that the omnidirectional emission pattern of the proposed antenna makes it appropriate for skin implantation. This suggested antenna performs better than earlier attempts in terms of size, bandwidth, and gain.

[64] Hassan T Chattha et al., provides applications for 4G, 5G, and the Internet of Things (IoT) a versatile wideband multi-input multi-output antenna with excellent port-to-port isolation. Two identical planar inverted-F antennas are used in this antenna arrangement (PIFA). To attain wideband properties, each PIFA's top plate has closed-ended and open-ended rectangular slots carved out of it, and its ground plane also has a slot etched into it. For high separation between two PIFAs, the ground plane's centre has a rectangular slot carved out of it. In order to establish an antiresonance for enhancing isolation between two PIFA elements and further minimising mutual coupling, a thin rectangular strip is connected between the top plates of two PIFAs.

[65] Narinder Sharma et al., discusses the construction and investigation of a nested hexagonal ring-shaped fractal antenna. Antenna 1 is initially a multiband, wideband integrated hexagonal concentric ring antenna. To better multiband/wideband behaviour, antennas 2, 3, and 4 have the patch orientation of antenna-1 adjusted. Comparing the performance of the designed antennas, it was found that antennas 2 and 3 outperform antennas 1 and 4. Antenna 2's bandwidth is 1.26 GHz (2.5–3.7 GHz), 2.75 (4.0–6.7 GHz), and 6.1 (7.8-13.9 GHz). Equivalently, antenna 3 has a maximum gain of 6.1 dB, a bandwidth of 340 MHz (1.9-2.2 GHz), 820 MHz (3.0-3.86 GHz), 4230 MHz (5.3-9.6 GHz), and 3040 MHz, and a frequency range of (10.4-13.4 GHz). To confirm the results, a sample of the suggested antennas with good alignments is constructed and tested. Also examined and found to be in good agreement are the results from both simulations and measurements. In the E-plane and H-plane, antennas exhibit bidirectional and omnidirectional patterns, respectively. Designed antennas can be utilised for a range of wireless technologies since they have a greater bandwidth.

[66] Sreenath Reddy Thummaluru et al., designs a versatile reconfigurable filter with three working modes is provided. All-pass filter, adjustable bandpass filter, and tunable band reject filter are the three operational modes. The constructed filter was combined with an ultra-wideband antenna to create a reconfigurable filtenna. These four filtennas are stacked on a single substrate, and sufficient isolation is ensured between them by utilising reflectors to construct a four-port multiple-input multiple-output (MIMO) filtenna suitable for Mid-band 5G cognitive radio (CR) applications. The finalized designed structure provides sensing and communication antennas for spectral range and frequency band CR by changing the operational modes of the filter. By doing the measurement, the simulated findings were validated. The proposed antenna's MIMO performance was assessed by measuring ECC and found to be suitable for practical uses.

[67] Nand Kishore et al., describes a unique SWB antenna design is shown that generates a wide operational bandwidth without altering the ideal size of the antenna by substituting inverted double U-slots on the ground plane for double U-slots on the radiating patch. The ground plane is widened by inserting a triangle incision. As a result, lessen the ground plane impacts. For both computational and experimental methodologies, the effects of slot and slit concepts on antenna performance are clarified. The recommended antenna is a viable option for SWB applications because to its extremely wideband bandwidth of 2.4 GHz to 32.9 GHz.

[68] Eunyoung Park et al., describes an overview of the study, demonstrates its design, and assesses its performance as a transverse electromagnetic mode rectangular coaxial waveguide for a mobile 5G milli-metre wave antenna module. The general architectures of 5G radio frequency modules for mobile sub-6 GHz and mm - Wave applications are also covered. For building fully integrated RF front-end modules, thorough assessment of transmission line and device topologies in the mm-Wave frequency band is essential. The rectangular waveguide structure's impedance value was determined using the geometric mapping method. Additionally given are a design guide and a theoretical operation frequency bandwidth. The rectangular coaxial waveguide structure evaluated in this work and its parameters are easily adaptable to different frequency bands.

[69] He Huang et al., delivers a remarkably low-profile, single-ended, dual-polarized patch antenna. It consists of three substrates, each of which is responsible for feeding, transmitting energy, and transmitting transmission. A patch is printed on the upper substrate's top layer. The patch has four triangular slots carved into it to carry current in a slant direction and lessen the impediment it causes to nearby structures. Then, to increase structural symmetry and link the energy to the patch, two coupled lines are put crosswise on the top layer of the medium substrate. The linked lines are connected to the transition lines on the bottom side of the lowermost substrate using plated through holes.

[70] Melvin Chamakalayil Jose et al., delivers a special compact, ultra-wideband operationally switchable dual-band microstrip planar printed rectangular monopole antenna with omnidirectional to directed radiation pattern. Two symmetrical grounds, two reflectors, and a combination of the three are arranged on the same layer of the rectangular microstrip PRMA. To achieve frequency flexibility and a switch from an omnidirectional to a focused radiation pattern, two SMD PIN diodes are required. Directed radiation patterns with 180° phase transitions are made possible in the C-band frequency range. The suggested antenna system's numerous design parameters were parametrically evaluated, and the antenna's properties were investigated. The printed circuit board etching technique is used to fabricate an antenna prototype utilising RMI UV laser etching and cutting tools. The suggested antenna properties are appropriate for high-speed wireless sensor network connection in industries and pharmaceuticals. It's also ideal for IoT and 5G sub-6-GHz applications.

[71] Aabia Khan et al., shows how a fabric substrate was used to construct a reconfigurable wearable antenna. The developed antenna might work with a frequency change to operate in the 5G mid band. The suggested antenna can be used for Wi-Fi, WLAN, satellite, 5G mobile and cognitive radio communication and can be carried as a part of a uniform or any other type of apparel. By combining all the mentioned reconfiguration elements into a single antenna and performing a parametric analysis, the antenna shape was created. This may enable successful results when using fabric like polyester as the substrate.

2.2 A REVIEW ON DIFFERENT MIMO ANTENNAS

[72] Sachin Kumar et al., outlines an innovative design approach for a circularly polarised rectangular MIMO antenna with an ultra-wideband. The four-unit cell antennas, microstrip feed line, and square slotted ground plane make up the intended MIMO antenna. In the suggested unit cell architecture, circular polarisation is attained via a circular stub projecting from the ground plane strip. By changing design parameters, the unit cell of the MIMO antenna is maximized. The proposed four-port MIMO antenna design has an impedance bandwidth of 112% (3.1-11 GHz) and a 36% axial ratio bandwidth. The compact four-port MIMO antenna prototype is constructed on a FR4 substrate (4.8–6.9 GHz).

[73] Niraj Kumar et al., designs the creation of a one-of-a-kind, small EBG structure and its application to enhancing isolation in the H-Plane of a WLAN MIMO antenna system. An isolation or coupling decrease of 5.6 dB is achieved by etching the proposed EBG structure off the ground plane of a microstrip patch MIMO antenna. EBG is compact, which decreases the centre-to-centre distance. To further decrease mutual coupling at 5.8 GHz, radiating patches are connected with a metal line strip. With the addition of metal line strip, isolation is significantly improved by 16.2 dB. This results in a total mutual coupling reduction of 21.8 dB, which is further supported by the surface current plot.

[74] Charles MacWright Thomas et al., provides a research study on MIMO communication systems using a V-shaped microstrip patch antenna. Researchers compute and simulate the 2.45 GHz microstrip patch antenna initially. Then, a variety of MIMO antenna components are simulated and discussed. The V-shape with a 45° inward and outward tilt is being investigated. A small antenna design is chosen after considering the antenna properties and the simulation results. The findings demonstrate that low mutual coupling may be maintained with a 1 mm reduction in the distance between antennas.

[75] Alaa M Hediya et al., suggested a 6-element MIMO antenna setup for use in the 5G communication range. The proposed antenna element is a four-section connected line folding antenna. By integrating it with the DGS in between the various components of the MIMO antenna, this antenna element's efficiency is increased. In this case, the separated single antenna spans the frequency range of 3 GHz to 5 GHz with a reflection coefficient of less than 10 dB. All antenna elements in the MIMO system have a reflection coefficient that is less than 7 dB in the same frequency band. However, in the MIMO design, the distance between antenna components is greater than 15 dB.

[76] Leeladhar Malviya et al., delivers a compact 2x2 dual-band MIMO antenna with polarisation diversity for current wireless applications. The recommended design combines the radiating components that are both horizontally and vertically polarised. Positioning orthogonal antenna elements and slightly stepped ground reduces mutual interaction between radiating elements. The frequency ranges between 2.4 and 2.7 GHz and 4.9 and 5.6 GHz are tested with SWR 2. More than 21 dB of isolation between adjacent and diagonal ports has been reported. Peak gains of 3.9 dBi and 4.1 dBi, respectively, were obtained at the resonant frequencies of 2.5 GHz and 5.2 GHz.

[77] Muhammad Jaafer Riaz et al., describes about the introduction of MIMO antennas for future 5G Applications. These channels are reserved for 5G communications, and the proposed effort is to build a MIMO antenna that functions in one of the FCC-reserved 5G communications frequencies. Because MIMO technology already supports high transmission rates, an increase in data traffic can be addressed by the MIMO system. The suggested design is made up of seven parts, each of which operates at 37 GHz. The MIMO system offers better isolation (more than 20 dB) and a single element gain of 7.7 dBi, indicating that this suggested MIMO network is an excellent contender for future communications.

[78] Amjad Iqbal et al., describes how a modified U-shaped resonator improves the isolation of two densely grouped multiple-input multiple-output (MIMO) antenna systems. Two tightly packed radiating elements with a 1 mm edge-to-edge spacing and a 5.4 GHz resonance surround the modified U-shaped resonator. Rigid parametric modelling tuning is used to obtain the isolation level of 23 dB between closely spaced sections. By carefully creating the relevant circuit model, the coupling behaviour of the MIMO elements is studied in each phase. Apart from mutual coupling, the results show that there are very negligible effects on the antenna performance when decoupling structures are present or not. The simple assembly of the suggested modified U-shaped isolating structure makes it suitable for a wide range of related applications.

[79] Huey Shin Wong et al., provides a printed double-L sleeve monopole antenna and a printed microstrip antenna that work as a multiple-input multiple-output (MIMO) antenna for the LTE 1800 wireless application. Power for the printed double-L sleeved monopole antenna comes from a 50Ω coplanar waveguide (CPW). A new T-shaped microstrip feedline that is etched on the opposite side of the PCB is utilised to excite the waveguide's outer shell. The proposed MIMO antenna provides isolation qualities that are superior to 15 dB. The proposed antenna supports LTE 1800. (1710 MHz–1880 MHz). The omnidirectional characteristics of this antenna. The antenna has a high gain of 2.18 dBi and is more than 70% efficiency.

[80] Xiongwen Zhao et al., delivers a special dual-band frequency reconfigurable MIMO patch-slot antenna design based on microstrip feedlines. Four rectangular patch antenna elements in symmetrical alignment make up the plane antenna. A defective ground structure (DGS) with a dual purpose for reducing antenna size and improving isolation is created in the ground plane. The antenna is printed on FR4 substrate and has a modest overall footprint, measuring 120x60x1.6 mm^3. In the feedline, varactor diodes are used to achieve frequency reconfigurability. The recommended antenna design offers an overall isolation of more than 12 dB and a broad continuous dual frequency reconfigurable characteristic ranging from 1.3 to 2.6 GHz.

[81] Aftab Ahmad Khan et al., developed a novel dual-band multiple-input multiple-output (MIMO) rectangular dielectric resonator antenna (DRA) for WiMAX (3.4-3.7 GHz) and WLAN (5.15-5.35 GHz) applications. A compact design is created by layering a high permittivity material. The obtained impedance bandwidth is 9.9% at 3.6 GHz and 8.8% at 5.2 GHz. At 3.6 GHz, the antenna gains obtained from the two ports are 5.7 dBi and 6.6 dBi, respectively. Co- and cross-polarisation, radiation properties, antenna gain, envelope correlation, and mean effective gain of the proposed design are all measured. According to the findings, the suggested design is suitable for MIMO WiMAX/WLAN applications.

[82] Yixin Li et al., offers a 12-port antenna array for 5G MIMO applications in mobile devices that operates in the long-term evolution (LTE) bands as 42 (3400-3600 MHz), 43 (3600-3800 MHz), and 46 (5150-5925 MHz). The following is the suggested MIMO antenna: It primarily comprises of three different types of inverted Π-shaped, longer inverted, and shorter inverted L-shaped open slotted antennas. There are a total of eight antenna components. It displays improved isolation. Additionally, the suggested antenna array displayed good MIMO performance in LTE bands 42/43 and 46, respectively, with an envelope correlation coefficient less than 0.15 and average channel capacity better than 34 and 26.5 b/s/Hz. Additionally, the hand phantom effects are investigated, and the results show that the suggested antenna array may still perform well in terms of radiation and MIMO when operating in data mode and read mode.

[83] Gourab Das et al., explains a circular dielectric resonator (CDR)-based multiple-input multiple-output (MIMO) antenna system for 4G applications. The introduction of orthogonal modes in CDR has improved the isolation of the proposed antenna. Ports 1 and 2's respective relative impedance bandwidths are 11.9% and 10.6%, and the recommended radiator achieves a high isolation (S12) of more than -25 dB within the operational band. ECC, DG, and CCL are a few of the MIMO performance characteristics that are estimated and found to be within acceptable limits.

[84] Jie- Huang Huang et al., demonstrates a very well-isolated dual-band Multiple Input Multiple Output (MIMO) antenna system. This system makes use of two dual-band monopole antennas and a balancing transmission line. Each antenna element's operational frequency band lies between 5.2 GHz and 6 GHz and between 2.4 GHz and 2.6 GHz. To improve the isolation between these two antenna components, a completely decoupled transmission line is used. These two antennas estimated return losses in the operating frequency band exceed 10 dB. The measured isolation between the two antennas is higher than 15 dB. The operating frequency spectrum's overall envelope correlation coefficient (ECC) is less than 0.01. WiMAX, Bluetooth, and wireless local area networks (WLAN) can all be used with this configuration systems.

[85] Irene Kong Cheh Lin et al., features a triple band hybrid approaches were used to build a Multiple-Input Multiple-Output (MIMO) Rectangular Dielectric Resonator Antenna (RDRA) for Long Term Evolution (LTE) applications. The proposed MIMO antenna enables simultaneous data transmission and reception in LTE Band 8 at 0.9 GHz, LTE Band 3 at 1.8 GHz, and LTE Band 40 at 2.3 GHz. In this design, a meander line antenna and an RDRA are combined in a hybrid manner to enable multiband functionality. A meander line antenna has been proposed at 0.9 GHz across the lengthy vertical microstrip feeding line, whereas two modes of RDRA are used in this design, to apply a reduction in antenna size.

[86] Adham M Saleh et al., explores a low-profile dual-band (2x2) MIMO antenna that can use the 2.4 GHz, 5.2 GHz, and 5.8 GHz frequencies for WLAN applications. By placing a strip line between the radiating elements at the isolation frequency, the neutralising line approach is utilised to improve isolation (2.4 GHz). The antenna is fabricated and tested to verify the modelling results. The results in real life and simulations generally agree. The antenna achieves impedance bandwidths of 2.1 to 2.5 GHz and 4.5 to 6.4 GHz for reflection coefficient |S11| -10 dB. S21 results, on the other hand, have attained - 25 dB at 2.4 GHz and - 35 dB around 5.5 GHz, and are better than -15 dB for the full band (1-6 GHz). There is a report and analysis of every MIMO antenna parameter.

[87] Tripta Kumari et al., generates a cylindrical dielectric resonator antenna for WLAN and WiMAX applications that is dual band multiple-input multiple-output. It consists of two radiators that are equally spaced and perpendicular. Each radiator is stimulated by a microstrip line that passes through a tiny rectangular slot. For more mode excitation, the suggested framework uses dual segment DRA, which appears to have superimposed geometry. A higher order mode is produced by feeding the upper CDRA from the aperture fed dielectric resonator. The frequency range in which the stated radiator operates is from 3.3-3.8 GHz to 5-5.7 GHz. The measured isolation is higher than 20 dB in the required band. The suggested antenna achieves an average gain and radiation efficiency of 6.0 dBi and 85%, respectively, at the operating frequency band.

[88] Ming-Yang Li et al., provides a dual-polarized hybrid eight-antenna array for 5G communication with multi-input multi-output operation in a smartphone that operates in the 2.6 GHz band. The proposed hybrid antenna array elements consist of two separate four-antenna array types with orthogonal polarisation and are symmetrically positioned along the long edges of the smartphone. As a result, coupling between the two types of antenna arrays can be reduced, and the performance of MIMO systems can be enhanced. The prototype of the eight-antenna array is constructed and put to the test. It is feasible to obtain satisfactory isolation (better than 12.5 dB), adequate cross-polarization discrimination (better than 15 dB), and adequate impedance matching (10 dB return loss or better). The performance of the suggested antenna for MIMO is also examined along with the envelope correlation coefficient and channel capacity.

[89] Assad Iqbal et al., discusses an innovative single element dielectric resonator antenna (DRA) for the sub-6 GHz band with four port MIMO. A coplanar waveguide (CPW) supplies the DRA, which is shaped like a symmetric cross and is developed and constructed in that manner. Rarely a single radiator with four ports is mentioned in the literature. The antenna covers fifth generation (5G) new radio (NR) N47 bands and wireless local area network (WLAN) IEEE 802.11a bands with a -10 dB impedance bandwidth of 5.52 GHz to 6.2 GHz (11.6%). While the isolation between opposing ports is only about 12 dB, it is approximately 15 dB between orthogonal ports. The proposed antenna has a bidirectional radiation pattern because there isn't a ground plane underneath the DRA.

[90] Leeladhar Malviya et al., illustrates the utilisation of MIMO antenna arrays in 5G wireless technology. In order for the losses to be as low as possible at higher frequencies, it needs a substrate with a low loss tangent. Here, a Rogers RT Duroid 5880 with a 2x8 array is suggested for the design. The substrate's dimensions are 23.61 x 55.18 mm^2. Each patch has an inset-feed that is coupled to a 100-ohm microstrip line and is used for impedance matching. Gain and efficiency are both high and well suited to outdoor settings here.

[91] M V Karthikeyan et al., highlights how the cornerstone of current wireless technologies is the multiple input multiple output (MIMO) antenna. A number of mutual coupling reduction techniques are required for MIMO antenna design in a confined location; otherwise, gain, efficiency, diversity gain, and radiation patterns will suffer. Numerous strategies for reducing this degrading factor and enhancing MIMO antenna performance have been documented in the literature. In this review work, we conducted a thorough analysis of diversity and mutual coupling (correlation) reduction strategies in small MIMO antennas.

[92] Nikesh Kumar Sahu et al., develops a hybrid triple-band dielectric resonator with numerous inputs and outputs that is based on this theory. The proposed MIMO antenna consists of two symmetrical modified Y-shaped microstrip printed lines and two cylindrical dielectric resonators. Two metallic strips and a slit in the ground plane are used to increase isolation. A prototype of the suggested design is created and tested in order to validate the simulation results. With fractional bandwidths of 34.4%, 14.2%, and 14%, respectively, the measured findings showed that the recommended antenna can operate in three distinct frequency bands: 2.2-3.1 GHz, 3.4-3.9 GHz, and 5.3-6.1 GHz. In order to obtain circular polarisation in the upper frequency band with a bandwidth of 4.18%, a modified Y-shaped printed line is used.

[93] Su Xu et al., presents a novel metamaterial-based decoupling strategy with an extensive relevance for MIMO antennas in mobile phones. Subwavelength metal/air layers that can be viewed as a single medium over a broad frequency range make up the decoupling element. The same metamaterial decoupling device is used to demonstrate the decoupling strategy's adaptability to other antennas and frequency bands. Finally, it has been demonstrated that a 24-element MIMO antenna in mobile devices can provide 1/100-wavelength 10-dB isolation over the whole frequency range of 4.55 to 4.75 GHz.

[94] Naser Ojaroudi Parchin et al., creates a multi-band slot antenna array architecture that works with 4G and 5G handsets. The system consists of double-element square-ring slot radiators that are supplied by microstrip-line architectures, making it simple to interface with RF/microwave circuits. The smartphone printed circuit board is where the slot radiators are located. The multiple-input multiple-output (MIMO) antenna that has been proposed is meant to meet the bandwidth requirements of 4G and 5G mobile terminals for higher data rate applications. For a 10 dB impedance bandwidth, each single-element of the proposed MIMO design may cover the frequency ranges of 2.5-2.7 GHz (LTE 2600), 3.45-3.8 GHz (LTE bands 42/43) and 5.00-5.45 GHz (LTE band 46). Examining the specific absorption rate (SAR) levels of the antenna and the model findings for user-hand/user-head impacts and good results are obtained. Furthermore, the antenna elements can be employed as 8-element/dual-polarized resonators.

[95] Pawan Kumar et al., provides a tiny, low-profile quad-port multiple-input multiple-output (MIMO)/diversity antenna with triple band-notched (Wi-MAX, WLAN, and X-band) capabilities for super-wideband (SWB) applications. The proposed concept activates four equivalent truncated-semi-elliptical-self-complementary radiating patches through tapered CPW feed lines. In the ground plane of the truncated semi-elliptical antenna element, a complementary slot matching the radiating patch is positioned in order to obtain SWB. A range of 1.3 to 40 GHz is covered by the impedance bandwidth ($|S11|$<- 10 dB) of the proposed MIMO/diversity antenna. In addition, a complementary split-ring resonator is inserted in the resonating patch to suppress SWB from WLAN and X-band broadcasts. In order to remove interferences from the Wi-MAX band, an L-shaped slit is also used.

[96] Naveen Jaglan et al., presents a MIMO/Diversity fourth/fifth generation (4G/5G) suitable 18-element antenna system for mobile devices. The antennas are appropriate for sub-6 GHz Long Term Evolution (LTE) bands 42 and 43 (3.4-3.6 GHz, 3.6-3.8 GHz) respectively. Simple open-ended slots are used in the radiating element's design to provide a compact structure. These slots also act as decoupling elements, enhancing radiator isolation. The suggested antenna elements are made for 6-inch smartphones because they fit on a low-cost FR-4 substrate that is 150 mm x 80 mm x 1.6 mm in size. The values of the measured and simulated antenna gain are both more than 5.3 dBi.

[97] Rifaqat Hussain et al., provides a 4-element dual-mode meandered F-shape MIMO antenna that is reconfigurable for multiple-input multiple-output. To achieve frequency reconfigurability, a combination of varactor and PIN diodes is utilized. Given that, it covers a broad range of frequency bands below 1 GHz, the suggested architecture is advantageous. The proposed concept is portable and lightweight, making it perfect for usage in wireless handheld devices powered by cognitive radio. The isolation between different antenna elements was improved by a flawed ground plane design. The bandwidth of the two modes is at least 60 MHz.

[98] Sharjeel Riaz et al., uses a new rectangular slot-based MIMO frequency-reconfigurable antenna design for cognitive radio front-end applications. Eight rectangular antenna elements that are engraved onto the ground plane make up this antenna. In order to provide multi-band operation, the suggested design uses a meandering type slot in the centre of each antenna. In addition to two extra bands at 3.6 and 5.2 GHz, it offers a frequency tunable band between 1.6 and 2.4 GHz that covers a wide range of wireless protocols. Varactor diodes are placed in the middle of each element to achieve frequency reconfigurability by changing the capacitive reactance of the slot. Furthermore, to encourage isolation between antenna pieces that are positioned vertically, rectangular defective ground slots are utilised. On a 120x60x0.8-inch surface, the full pattern is realized.

[99] Peng Yang et al., introduces a tiny dual-band antenna array for MIMO (multiple-input multiple -output) technology on smartphones. The smartphone's long side edge frame is where the antenna elements of the array are mounted. The antenna is designed to operate in the fifth-generation (5G) band and has a planar shape and a modest footprint of 13.8 x 7 mm^2 (3400-3600 and 4800-5000 MHz). Two symmetric four-antenna arrays that run down the smartphone's two long side edge frames make up the whole array. A 2 mm separation is achieved between adjacent antennas in each four-antenna array. The antennas' envelope correlation coefficients (ECC) in the operational band are less than 0.1, and their isolation is satisfactory.

[100] Haleh Jahanbakhsh Basherlou et al., suggests a compact multiple-input, multiple output (MIMO) monopole antenna for 5G Wi-Fi applications. The antenna is constructed on an affordable FR-4 substrate with features. The proposed design has erroneous T-shaped ground planes and circular ring radiators that are microstrip-fed (DGS). By incorporating the T-shaped DGS, the antenna's mutual coupling characteristic has been reduced, and good isolation has been achieved. The antenna covers the required frequency bandwidth for 5G Wi-Fi networks by operating in the 5 to 6 GHz frequency range, with a reflection coefficient of 30 dB and mutual coupling of 20 dB. Analysed and presented are the simulation results for various antenna specifications.

[101] Raed A. Abd-Alhameed et al., describes a multiband multiple-input multiple-output (MIMO) antenna to meet this requirement for different mobile terminals. The multi-mode loop radiator and two connected parasitic structures that are positioned at the upper edge of the smartphone printed circuit board make up the antenna element of the 2x2 MIMO design. The self-balancing loop antenna can operate in a variety of frequency bands. The substrate is a low-cost FR-4 laminate measuring 75x150x1.6 mm3 in total. The antenna is easy to construct and install to its straightforward, compact, and flexible design.

[102] Atta Ullah et al., showcases a 6-element dual-band MIMO F-shape antenna for mobile communication using smartphones. It has six antenna elements that cover the dual-frequency sub-6 GHz bands of 3.4-3.7 GHz and 5.3-5.6 GHz. The ground is carved into a L- shape with each antenna element, while F-shaped microstrip feeding sites provide the basics. The core components of the suggested strategy are looked at. It is efficient, focused, and has the appropriate S-parameters, isolation, and radiation analysis. According to the dual-band frequency of 5G smartphones, Antenna 3 has a maximum return loss of -35 dB at 3.6 GHz and Antennas 5 and 6 have a maximum return loss of -38 dB at 5.4 GHz.

[103] Habib Ullah et al., shows a simple planar MIMO antenna for super wide band applications. Two identical patches put on the same substrate make up the stated MIMO antenna. The percent impedance bandwidth and bandwidth ratio of the SWB antenna, which was constructed on a substrate, where 187% and 30:1, respectively. Encasing a T-shaped corrugated strip between two antenna elements reduces mutual coupling between the elements. The proposed MIMO antenna offers efficiency levels above 80%, a maximum diversity gain of 10 dB, low mutual coupling (20 dB), a low envelope correlation coefficient (ECC < 0.02), and a low reflection coefficient (-10 dB) in the SWB frequency range (1.30 GHz–40 GHz).

[104] Mohammad Alibakhshikenari et al., outlines a technique for enhancing the isolation between close-by radiating elements, a characteristic of highly compacted antenna arrays. These antennas offer the frequency beam-scanning capabilities needed in synthetic aperture radars and MIMO systems. The method presented here uses a metamaterial decoupling slab (MTM-DS) that is sandwiched between radiating elements to prevent mutual coupling, which would otherwise impair antenna performance and functionality in both transmit and receive modes. The envisioned MTM-DS is made up of mirror-imaged E-shaped slits carved onto a microstrip patch with an inductive stub. The method also avoids the poor front-to-back ratio seen in other decoupling approaches. Additionally, setting it up is quite simple. The MTM-DS can be retrofitted onto existing antenna arrays assuming adequate space between adjacent radiators, making the recommended technique adaptable.

[105] Haq Nawaz et al., implements the design and development of a wide-band dual port monopole antenna based on a single circular disc radiating element for Multi Input Multi Output (MIMO) configured 4G mobile and wireless communication systems. In order to achieve an improved antenna's impedance bandwidth of 2-6 GHz, the suggested antenna employs a partial ground plane with two rectangular slots that lay immediately below the corresponding 50Ω microstrip feeding lines. To eliminate port-to-port RF coupling, a 14 mm radius circular cut is etched and centred at the junction of two partial rectangular ground planes. The two- port monopole antenna developed on the RT/Duroid® 5880 toward such more than 15 dB RF inter-port isolation for the antenna's 10 dB impedance bandwidth of 2-6 GHz, with measured gain variations of 2-6 dBi for each port excitation. In addition to measuring inter-port isolation for dual port antennas, the envelope correlation coefficient throughout the required frequency range was calculated and plotted using S11, S22, and S21 measurement findings to validate the performance of dual port implemented antennas for diversity applications.

[106] Vigneswaran Dhasarathan et al., builds a prototype for a four-port dual notched bands multiple-input multiple-output (MIMO) antenna design. The creation process starts with a single instance, which is later modified to a 2x2 MIMO configuration, and finally to a 4x4 MIMO configuration. The proposed MIMO antenna will be beneficial for X-Band and ultrawideband applications because it is designed to cover an impedance bandwidth of 3.15–11.36 GHz. The use of an inverted T-shaped stub and a C-shaped slot on the MIMO antenna's radiating patch prevents two overlapping bands, WiMAX and WLAN. Applications requiring high-speed wireless communication can use the provided antenna.

[107] Ravindra Bakale et al., examines the creation of a low mutual coupling, dual-band-elimination ultra-wideband MIMO antenna. The suggested architecture includes a microstrip fed monopole antenna with a stub to enhance isolation for ultra-wideband applications. Its projected mushroom-type electromagnetic band-gap structures, which are to be installed adjacent to the intended antenna's microstrip transmission line, serve as the foundation for the dual rejection bands that correspond to WiMAX and WLAN frequencies. Two counter-facing F-pattern stubs can be connected to the ground to improve isolation by $|S21| > 20$ dB over the impedance bandwidth. An impedance bandwidth of 9 GHz (2.6-11.6 GHz) for VSWR 2.13 and the notch bands of 3.6-4.2 GHz and 5.1-5.8 GHz is discovered. With regard to wireless applications, this is very common.

[108] Muhammad K Khan et al., describes a new dual-band notched Ultra-Wide Band (UWB) Multiple Input Multiple Output (MIMO) antenna. The intended MIMO antenna features two similar Mushroom-shaped radiators with a total size of 18x34x1.6 mm^3. To improve port isolation, inverted L-structured stubs are linked to the antenna's ground. Over the 3.0-12.8 GHz frequency, the suggested antenna provides increased isolation of 23 dB. To introduce a notched WiMAX band (3.3-4.3) and WLAN (5.0- 5.8) GHz band, two innovative strips are extruded in the antenna's ground and mushroom-shaped radiator. The provided antenna's peak gain ranges from 2 to 4.8 dBi, and its radiation efficiency ranges from 78 to 90%, with the exception of the (3.3-4.3) GHz and (5.08-5.8) GHz stop bands.

[109] Hrishikesh Narayan Tripathi et al., gives a compact reduced ground plane printed antenna for WLAN/WiMAX applications. The lowered ground plane antenna is built on a FR-4 substrate with dimensions of 30 mm x 15 mm x 0.8 mm. The antenna measures 10 mm in length and 15 mm in breadth. The small antenna has a wide bandwidth and is suitable for WiMAX and WLAN Dongle applications.

[110] Phalguni Mathur et al., implements a multi-purpose dual Multiple-Input-Multiple-Output (MIMO) antenna system for 5G mobile systems. The suggested antenna is made up of two sets of MIMO antennas mounted on a single board. The board's 8-port MIMO antenna operates in the 3.5 GHz (3.4 GHz-3.6 GHz) 5G band and has dimensions of 150 mm x 70 mm. The 4-port MIMO system, which is mounted on the chassis, operates on the 5.2 GHz WLAN/5.5 GHz WiMAX/5.8 GHz/6 GHz Wi-Fi band and has dimensions of 20 mm x 7 mm. The general architecture of the dual MIMO system is fairly simple, allowing plenty of room on the board for other components. The availability of two different MIMO antennas in the proposed prototype allows for independent use of 5G and 4G frequencies. Because no active parts like as diodes, switches, and so on are used in the proposed design to transition between different bands, the prototype has no ohmic losses. It's also worth noting that the proposed antenna works in the much-anticipated next-generation Wi-Fi 6E band.

[111] Libin Sun et al., suggests a new self-decoupled multiple-input multiple-output (MIMO) antenna pair with shared radiator for fifth-generation (5G) cell phones. In our technique, a radiator is directly excited by two feeding ports, which are inherently isolated across a large bandwidth without the use of any additional decoupling structures. For the first time, a mode cancellation method based on the synthesis of common and differential modes is created to provide a detailed physical understanding of the self-decoupling mechanism. The proposed self-decoupled antenna pair exhibits greater than 11.5 dB isolation across the 5G N77 band (3.3-4.2 GHz) with radiation pattern variety. To test the concept, a MIMO antenna system comprised of four sets of antenna pairs is simulated, built, and measured using the self-decoupled antenna pair. The experimental results show that the proposed MIMO system can provide better than 10.5 dB isolation between all ports and a high total efficiency of 63.1-85.1% spanning 3.3-4.2 GHz. The suggested design concept, with its advantages of self-decoupling, shared radiator, simple structure, broad bandwidth, and high efficiency, shows remarkable potential for future strongly MIMO antennas for 5G handsets.

[112] Anping Zhao et al., proposes to use a broadband 8-antenna multiple-input and multiple-output (MIMO) technology in mobile terminals for 5G new radio N77 (3.3-4.2 GHz). Each antenna component, in particular, is made up of a monopole-type meandering branch and an inverted U-shaped self-isolated antenna branch, and the antenna unit is fed by the meandered division. The antenna unit has two resonances, one for each branch of the antenna unit. Because the monopole links strongly with the self-isolated antenna element, which has intrinsic self-isolation properties, the broadband 8-antenna MIMO system can achieve relatively good isolation. The results include antenna efficiency, radiation patterns, the envelope correlation coefficient, and channel capacity.

[113] Wei Hu et al., implements a dual-band eight-element MIMO array for fifth generation (5G) mobile communication that employs a multi-slot decoupling approach. The suggested array achieves two broad bandwidths of 12.2% and 15.4% for sub-6 GHz operation by employing a tiny dual-loop antenna element. A unique dual band decoupling approach utilising a multi-slot structure is developed to reduce mutual coupling between antenna elements. Across the two operational bands, the proposed MIMO array achieves 15.5 dB and 19.0 dB isolation. In addition, three decoupling modes caused by various bent slots can be tweaked independently. The coplanar layout of the antenna elements and decoupling devices also allows for zero ground clearance. The suggested MIMO array was designed, built, and tested. The experimental and computational results agree well, demonstrating that the dual-band MIMO array has good impedance matching, high separation, and highly efficient. In addition, the envelope correlation coefficient and channel capacity are calculated and examined to validate the 5G terminal array's MIMO capability. A dual-band high-isolation eight-element MIMO array with zero ground clearance is an attractive choice for 5G or future mobile applications.

[114] Zhouyou Ren et al., outfits a dual-band four-element MIMO antenna system based on tiny self-decoupled antenna pairs for fifth-generation (5G) mobile terminal operation. Dual-band antenna pairs with good isolation can be created by sharing one common grounding branch for the two neighbouring antenna units. The two compact antenna pairs, in particular, are arranged perpendicularly on both sides of the system ground plane. The MIMO antenna system is designed to operate in both the 3.5 GHz and 4.9 GHz bands, with isolation more than 17.5 dB in the low band and 20 dB in the high band. The suggested dual-band four-antenna MIMO system is built and measured, and there is strong agreement between simulation and measurement. Additionally, the effects of the phantom hand and the display panel on the efficiency of the MIMO antenna system are investigated and addressed.

[115] Samir Salem Al-Bawri et al., establishes slot-based broadband single and MIMO antennas for 5G New Radio (5G NR) mobile applications. Putting two antennas onto a low-cost material with a partial ground for sub-6 GHz is used to investigate the proposed MIMO antenna. Each antenna element is made up of a hexagonal-shaped slot and a 50-ohm microstrip feed line. With good impedance matching, the WLAN 5 GHz spectrum and 5G NR Bands are covered. Furthermore, the mutual coupling between neighbouring elements is less than 15 dB, while the desired antenna element gains are 3.19 dBi and 3.09 dBi at 3.5 GHz and 4.2 GHz, respectively.

[116] Huanqing Zou et al., introduces a dual-functional antenna array in the long-term evolution (LTE) band and the wireless wide area network (WLAN) band for multiple input multiple output (MIMO) applications in 5G handsets. The suggested eight-antenna array is made up of two distinct antenna elements: the folded L-shaped antenna and the couple-fed U-shaped antenna. It can be used in the 8x8 (using 8 antenna elements) and 4x4 (using 4 antenna elements) MIMO systems, respectively, across the LTE band 42 and the 5.8 GHz WLAN band. To obtain high isolation within the operation bands, quasi-orthogonal polarisation, balanced mode excitation, and pattern diversity are used effectively. Furthermore, the effects of the user's hand and head are explored, and desirable findings are produced. The results presented above indicate that the suggested antenna array is a promising contender for MIMO applications in smartphones.

[117] Xuekang Liu et al., explains a dual-band ten-element MIMO array based on dual-mode inverted-F antennas (IFAs) for 5G terminal applications. The suggested dual-mode IFA is made up of two radiators engraved on the outside and inside surfaces of the side-edge frame. The low-order mode is generated by the antenna's outer portion at 3.5 GHz, while the inner portion emits another one-quarter wavelength mode at 4.9 GHz. As a result, the IFA may operate in dual bands while being tiny at 10.6x5.3x0.8 mm^3. Based on the suggested antenna, a dual-band ten-element multiple-input and multiple-output (MIMO) array for 5G terminal applications is being developed. The isolations between the elements are improved by combining neutralisation line structures with decoupling branches.

[118] Aidi Ren et al., introduces a new design strategy for improving the isolation of a two-antenna construction block. In this case, a two-antenna construction block (Antenna 1 and Antenna 2) made up of a gap-coupled loop antenna (Antenna 1) and a loop antenna (Antenna 2) is precisely built to allow Antenna 1 to excite a standing wave zone that can be formed on Antenna 2's structure. To achieve strong isolation (26 dB) between the two antennas, the stable current null point in the standing wave domain is used and engineered to fall at Antenna 2's feed point during Antenna 1's excitation. The proposed four- and eight-antenna MIMO arrays may provide desirable measured isolation of more than 23 dB and 17.9 dB, respectively, over the band of interest, with measured ECCs (envelope correlation coefficients) less than 0.032 and 0.075. The periodic channel capacities of the provided MIMO arrays are further explored to further assess their MIMO performances.

[119] Raj Kumar et al., describes a small Ultra-Wideband (UWB) antenna system for multiple-input multiple-output (MIMO)/diversity applications. The antenna system is a microstrip configuration with two radiating elements of comparable triangular form and a common ground plane. For improved impedance matching, the radiating elements are activated via a tapered feed. To improve separation between the two radiators, a T-shaped stub is inserted in the ground plane. The suggested antenna system has an overall diameter of 46 mm x 32.6 mm x 1.6 mm. The observed impedance bandwidth (reflection coefficient< -10 dB) spans the whole UWB area, from 1.8 GHz to well beyond 12 GHz. The UWB region achieves verified isolation of 15 dB or better.

[120] S Jaisiva et al., introduces a multiple band antenna. It is made up of two elements, which represent the Multiple Input and Multiple Output (MIMO) technology. It acts as an isolating device, reducing mutual coupling between the ports on it. It provides adequate isolation and return losses of 40 dB and 17 dB, correspondingly. It has a small planer profile with a spectrum of 1-3 GHz. It has given good Quality of Service (QoS) with no deterioration in bandwidth or transmission power. As a result, the simulation results verified these design parameters.

[121] Qurrat-Ul-Ain Nadeem et al., discusses full-dimension multiple-input multiple-output technology, which is undergoing major development and standardisation in wireless communications in preparing for the transition to Fifth Generation (5G) cellular systems. It employs an active antenna system (AAS) with a 2-D planar array structure, which not only enables a significant number of antenna elements to be packed into practical base station form factors, but also enables adaptive electronic beamforming in three - dimensional space. The small construction of large-scale planar arrays, on the other hand, significantly boosts correlation coefficients in FD-MIMO systems. To account for its effects, the AAS's generalised spatial correlation functions for channels composed of individual elements and overall antenna ports are constructed. The challenge of calculating the ideal downtilt weight vector for antenna ports, which maximises the least signal-to-interference ratio of a multi-user multiple-input-single-output system, is described as a fractional optimization problem using users' quasi-static channel covariance matrices. Finally, a user-group-specific elevation beamforming scenario is developed, which provides significant performance increases as demonstrated by simulations. These findings are directly applicable to the examination of 5G FD-MIMO systems.

[122] Rifaqat Hussain et al., introduces a unique, small, ultra-wideband tunable concentric pentagonal slot-line-based frequency reconfigurable multiple input-multiple-output (MIMO) antenna system. It is made out of a 4-element antenna on a widely available FR-4 substrate. All of the antenna elements have a plane construction and are carved from the ground (GND) plane. By reactively feeding the antennas, varactor diodes provide frequency reconfigurability. Each varactor diode capacitance value leads in triband functioning. The antenna may resonate over a wide frequency spectrum due to proper capacitive reactance loading. The suggested MIMO antenna system is adjusted over an ultra-wide frequency band, and the resonance frequencies vary smoothly between 1.32-1.49 GHz and 1.75-5.2 GHz. Furthermore, the effectiveness of reactive loading the slot throughout all resonant bands distinguishes the offered antenna system. The presented antenna system is small and lightweight, making it ideal for usage in wireless portable devices and mobile terminals for cognitive radio (CR) applications.

[123] Mohammad S Sharawi et al., proposes a revolutionary integrated antenna solution for wireless handheld devices. The full antenna system is a unique combination of a MIMO antenna system at microwave frequencies and a mm-wave antenna array. The MIMO antenna system is made up of two reactive loaded monopoles, whilst the mm-wave array is made up of planar 2 by 4 slot antennas. The integrated antenna system spans the frequency ranges for 4G standards from 1870 to 2530 MHz, as well as the forthcoming 5G mm-wave band at 28 GHz. Furthermore, the integrated antenna system is planar and is intended for common smart phone handsets with a 60 mm by 100 mm by 0.965 mm back plane. Excellent field correlation values were attained over the 4G band, with realised peak gain values of 4 and 8 dBi observed for the MIMO and mm-wave antenna arrays, respectively. The antenna design could also be beneficial for future compact 4G and 5G implementations.

[124] Le Chang et al., explains a polarization-orthogonal co-frequency dual antenna pair appropriate for a fifth-generation (5G) MIMO smartphone with metallic bezels. The spatial-reuse dual antenna pair proposed consists of a split on the metallic bezel and a slot on the mainboard ground, with the slot centred on the split. On the same antenna configuration, two orthogonal degenerated characteristic modes functioning at half wavelength, including in-phase current and slot modes, can be excited. A geometric slot-centered Y-shaped feeding network with shorted ends crossing the slot is used to excite the in-phase current mode, producing y-polarized radiation, while a slot-centered Y-shaped feeding network with shorted ends crossing the slot is used to excite the slot mode, producing x-polarized radiation. To test the practicality, a four-antenna MIMO array is built and measured. The measured isolation of the antenna pair is more than 20.1 dB over the operation band, and other isolation levels are all greater than 12.7 dB, the ECCs of any two antennas are less than 0.13, and the total efficiency range of the four antennas is 35.2%-64.7%. The proposed dual antenna pair has a high potential for use in future 5G MIMO smartphones.

[125] Igor Syrytsin et al., gives the performance of a beam-steerable antenna system at 3.5 GHz with user effects. An antenna array system in a metal structural frame is seen as an example to demonstrate how to performance analysis of such a system. Four phased subarrays are part of the antenna system. Every subarray has two slot antenna elements. Depending on the input power and interference directions, the system allows you to control the beam by phase in each subarray and swap the primary beam direction among the four phased subarrays. The basic idea is to monitor the performance of the phased array system using CTIA standards for total scan arrangement and coverage efficiency. In free space and with user impacts, the proposed antenna system is modelled and measured. It also includes and discusses user-induced scan angle and coverage efficiency losses.

[126] Chenyin Yu et al., implements a unique super-wideband multiple-input multiple-output antenna system with high isolation, with each antenna element composed of a spade-shaped radiation patch, a defected ground structure (DGS), and a windmill-shaped decoupling structure. The radiation patch's elliptical slot and the rectangular DGS operate together to improve radiation performance and impedance bandwidth. Four-element and eight-element SWB-MIMO antenna systems are both simulated and studied. The suggested SWB-MIMO antenna system may operate in the frequency range of 2.9 to 40 GHz with a mutual coupling of less than -17 dB, according to measurement and modelling results. A low envelop correlation coefficient (ECC) of less than 0.04, a high multiplexing efficiency (mux) of more than -3.0 dB, a constant gain of up to 13.5 dBi, and quasi-omnidirectional radiation qualities have also been achieved. This design serves as a useful guideline for creating a super-wide MIMO antenna.

[127] Wenfei Yin et al., proposes an extremely tiny MIMO antenna for portable wireless UWB applications. To achieve polarisation variety, the suggested prototype comprises of four monopole antenna elements positioned perpendicularly. Furthermore, the space between the radiation element and the ground plane is designed to reduce mutual interaction between antenna elements. Furthermore, a matching stub has been connected to the feedline to ensure high frequency impedance matching. The suggested antenna has a bandwidth of 3-20 GHz and a strong isolation of more than 17 dB, according to both simulated and measured data. Furthermore, the suggested MIMO antenna has outstanding radiation characteristics and steady gain across the whole operating spectrum. The ECC (envelope correlation coefficient) is less than 0.1.

[128] Zhongliang Deng et al., highlights a unique ultra-wideband (UWB) printed antenna with quadruple band-notched characteristics. The quadruple band rejections are obtained by etching two inter - digital capacitance slots on either side of the ground plane, embedding a curled C-shaped slot on the circle patch, and incorporating a curled split-ring resonator on the antenna's backside. Because of their large inductance, interdigital inductance slots can achieve a narrower notched band than conventional designs, retaining some useful frequencies. The notch frequency and width can be controlled by adjusting the length and distance of the branch from the curled C-shaped slot and the curled split-ring resonator. The antenna can accurately filter out the narrow band signals.

[129] Tayyab Shabbir et al., gives a low-cost 16-port non-planar Multiple-Input-Multiple Output (MIMO) antenna system for future 5G applications. The non-planar MIMO antenna system is built on a polystyrene block with a 3D octagonal shape. MIMO elements are put on the eight sides of an octagonal form block, leaving the bottom and top faces of the polystyrene block vacant. A slotted microstrip patch with a stepped chamfered feed line and a defective ground plane makes up the single antenna element. Each MIMO element is built on a FR-4 substrate with dimensions of 22 mm x 20 mm to cover the frequency range of 3.35 GHz to 3.65 GHz for fifth-generation (5G) applications. The array elements are on the top layer, while the common linked ground plane and decoupling structure are on the bottom layer. The metamaterial-based decoupling structure provides more than 28 dB isolation for antenna elements grouped across and side-by-side. Furthermore, the simulated and measured MIMO performance metrics are acceptable. The suggested non-planar 3D-MIMO antenna system can be used for indoor localization systems as well as wireless personal area network applications in which many 5G devices are wirelessly linked to a central controller. Furthermore, for the non-planar MIMO antenna system, there is good agreement between simulated and measured findings.

[130] Reza Karimian et al., presents a novel microstrip slot antenna with parasitic elements. The proposed antenna is made up of a microstrip feed line, a ground plane with simple slots etched into it, and parasitic elements. The simulation results show that the antenna structure allows for independent frequency tuning. This antenna's two-element array configuration for MIMO applications is also examined. Both single and array configurations were built to compare simulation and measurement results. In terms of return loss, low mutual coupling, and compactness, the measurement result shows good radiation performance.

[131] Anil Kumar Gautam et al., describes the design of a dual-band-notched ultrawide band (UWB) MIMO antenna that is exceedingly small. The suggested antenna's major characteristics are its small size and strong isolation between the elements. The antenna's UWB performance is achieved by creating two identical triangular-shaped patches that are designed in opposition to one another and are linked by a tapered microstrip feedline. The proposed antenna's ground plane is made up of a funnel-shaped patch joined at the bottom by two triangles. Furthermore, two J-shape slits are engraved in the radiator to obtain band rejection characteristics in the wireless local area network spectrum and IEEE INSAT/Super-Extended C-band. The smallest dimension of the planned antenna is $26x15x1.6mm^3$. The MIMO antenna has a broad impedance bandwidth of 31.9 GHz from 3.1 to 35 GHz, according to the results. This antenna is a promising choice for portable UWB applications due to its low mutual coupling of < - 24 dB and ECC of < 0.2 across the frequency spectrum.

[132] Jianfeng Zhu et al., proposes a tri-band MIMO mobile antenna for WWAN and LTE applications. The given antenna is coupled-fed by an L-shaped driven strip monopole to form the tri-band antenna element. A long parallel folding strip printed adjacent to the upper border of the PIFA generates an additional mode in the lower band, allowing the antenna to cover the whole GSM 850 and GSM 900 bands. To improve port isolation, two n-shaped strips are put between the antenna elements. The suggested antenna has three bands for WWAN and LTE applications: the lower band of 0.81-0.97 GHz, the middle band of 1.36-2.80 GHz, and the top band of 3.40-3.75 GHz.

[133] Jun-Yi Jiang et al., presents a wideband eight-element multiple-input multiple-output antenna array for 5G smartphone applications. Each antenna is made up of a dual-arm meandering monopole radiating element, a double-stub tuner, and an open ground plane slot. Impedance matching is improved by using tuning stub microstrip lines. The single antenna element's operational bandwidth ranges from 3.2-6 GHz with three resonant frequencies. Without any isolation devices, the proposed MIMO antenna array has isolation greater than 10 dB over the whole operating band. Furthermore, the proposed MIMO array was built and tested. The measured results confirm that the MIMO antenna array has a broad 6-dB impedance bandwidth from 3.2 to 6 GHz, and that the isolations are all more than 10 dB.

[134] Jayshri Kulkarni et al., designs a single radiator in a four-port diversity MIMO system which are made up of four octagonal rings placed between the two opposite sides of a T-shaped conducting layer and encircled by inverted angular edge cut L-shaped and E-shaped structures. The radiators are put at the four edges of a smartphone, with a common ground in the centre, to form a four-element mobile MIMO antenna. The antenna is printed on a flexible polyamide substrate with dimensions of 70x145x 0.2 mm^3. All four radiators have an impedance bandwidth of (84.12%) 2.39 to 5.86 GHz. The tiny size of the radiators, combined with their arrangement, allows the proposed MIMO antenna to take up much less space while still allowing space for 2G/3G/4G antennae. The antenna placement results in self-isolation between antenna elements of greater than 17.5 dB in the desired operating bands. Furthermore, the flexible MIMO smartphone antenna achieves good diversity performances such as ECC of less than 0.05, DG of above 9.8 dB, MEG of 3.1 dB, CC of 21.50 bps/Hz, and TARC of less than 10 dB, in addition to a high efficiency of 85% and adequate gain above 4 dBi.

[135] Surentiran Padmanathan et al., highlights the design of a multiband MIMO antenna in the 5G sub-6 GHz band with frequency and radiation pattern reconfiguration capability. PIN diodes and DC biassing circuits are used to enable frequency and radiation pattern reconfiguration on the antenna, which consists of two planar inverted-F antenna (PIFA) elements. Both PIFA element 1 and PIFA element 2 achieve triband with bandwidth spanning from 220 MHz to 2330 MHz, ranging from 0.8 GHz to 6 GHz, encompassing cellular bands for GSM, UMTS, LTE, and 5G-NR bands at reflection coefficients less than 6 dB. Furthermore, at least 10 dB isolation and an envelope correlation coefficient of less than 0.3 between ports assure adequate MIMO diversity performance. At their separate resonant frequencies, PIFA elements 1 and 2 produced main lobe gains ranging from 1.06 dB to 2.97 dB, with total efficiencies ranging from 46.2% to 74.5% attained within the operational bandwidth. For both PIFA elements, the proportion of channel capacity attained over IID capacity ranges from 73.71% to 92.8%. Simulated antenna performance parameters coincided well with measurements, implying that 5G mobile terminals could benefit from reliable and constant data transmission.

[136] Xiao-Ting Yuan et al., presents a wideband multiple-input-multiple-output (MIMO) antenna for 5G devices. First, an inverted-T-shaped open-slot with planar inverted-F antenna (PIFA) is developed. Because of the multimode technology, the intended PIFA has a 78% bandwidth. Then, two such PIFAs are combined to form a PIFA-pair, resulting in a 2x2 sub-MIMO antenna with a similar bandwidth as the PIFA-element. Furthermore, the isolation between two PIFA-elements in a PIFA-pair is greater than 10 dB across the whole bandwidth. Finally, a 4X4 MIMO antenna made up of two PIFA-pairs is presented to span the 5G NR frequency bands n77/n78/n79 as well as LTE band 46. The simulated and measured data are presented to demonstrate the suggested MIMO antenna's performance.

[137] Ashfaq Ahmad et al., introduces a simple, compact, and low-profile MIMO antenna that operates at 29GHz with reduced mutual coupling between the antenna elements for future 5G communication. To achieve excellent isolation among the antenna radiators, the suggested design employs two radiating elements with slits in the radiators. The MIMO antenna has a compact structure and is the smallest in size when compared to prior 5G antennas. As a substrate material, Roger's 4350B laminate was used. At 29 GHz, we achieve a low mutual coupling of 36 dB, a low envelope correlation coefficient (ECC < 0.001), and a large diversity gain (DG > 9.8 dB). The S-parameters, diversity gain, radiation pattern, and envelope correlation of the proposed design are all studied.

[138] Mohammad M Fakharian et al., proposes a quad-element, two-port, high-gain, multiband MIMO planar antenna based on a log-periodic dipole array (LPDA) for Ku/K-band wireless communication systems. A single element antenna was built and then refined with a 50Ω microstrip feed-line with two orthogonal branches, resulting primarily in a broadside radiation pattern and improved diversity characteristics. The planned structure is printed on an RT-5880 substrate with a thickness of 1.57 mm for field measurements. The MIMO antenna's total substrate dimensions are 55 x 45 mm^2. Furthermore, without the usage of any decoupling structure, the suggested MIMO antenna achieves a peak gain of 4.2-10.7 dBi with a maximum element isolation of 23.5 dB. In addition, an examination of MIMO performance measures such as the envelope correlation coefficient (ECC) and mean effective gain (MEG) supports strong characteristics and field correlation performance over the functioning spectrum. The proposed architecture is an appropriate solution for multiband MIMO applications in Ku/K-bands for various wireless systems.

[139] Samir Salem Al-Bawri et al., highlights a slot-based broadband single and MIMO antennas for 5G New Radio (5G NR) mobile applications. The proposed MIMO antenna is investigated by inserting two antennas into a low-cost material with a partial ground for sub-6 GHz. Each antenna element is made up of a hexagonal slot and a 50-ohm microstrip feed line. With good impedance matching, the WLAN 5 GHz spectrum and the 5G NR Bands n77/n78/n79 are supported. Furthermore, the mutual coupling between neighbouring elements is less than 15 dB, although the desired antenna element gains are 3.19 dBi and 3.09 dBi at 3.5 GHz and 4.2 GHz, respectively.

[140] Demos Serghiou et al., operates a dual-band 8x8 MIMO antenna operating in the sub-6 GHz spectrum for future 5G MIMO smartphone applications. The design consists of a fully grounded plane with closely spaced orthogonal pairs of antennas arranged symmetrically along the smartphone's long sides and corners. The orthogonal pairs are linked by a 7.8 mm short neutral line to reduce mutual coupling in both bands. Each antenna element is a folded monopole with dimensions of 17.85 X 5 mm^2 and can operate in the low band at 3100-3850 MHz and the high band at 4800-6000 MHz ($|S11| < -10$ dB).

[141] Kumud R Jha et al., proposes an angularly symmetric, common radiator coplanar waveguide (CPW) fed four-port multiple-input multiple-output (MIMO) antenna using a RT duroid substrate, where L is the free space wavelength at the lowest operational frequency of 0.6 GHz. To cover the expanding wireless communication bands, the antenna has a -6 dB impedance bandwidth from 0.6-1.09 GHz, 2.6-3.4 GHz, and 4.2-7.0 GHz. It also has a -10 dB impedance bandwidth spanning 0.7-1.01 GHz, 2.6-3.18 GHz, 5.3-6.06 GHz, and 6.7-6.94 GHz. The design steps for improving operating bandwidth and isolation in the sub-1GHz bands are provided. At the simulated and observed frequencies, the antenna has a reasonable realised gain. It demonstrates pattern variety, which is beneficial for MIMO implementation. The observed data were used to calculate the envelope correlation coefficient (ECC), mean effective gain (MEG), and channel capacity of the antenna. Despite the short circuit size at fL, the ECC is 0.50 across the whole band. This antenna can find fresh uses in the emerging 0.6-1.09 GHz band, sub-6 GHz 5G near radio (NR), and Wi-Fi 6 communications, in addition to existing communication applications.

[142] Zhouyou Ren et al., presents a small duplex structure with both sub-6 GHz and millimeter-wave antennas. The proposed antenna system includes two 1x8-antenna arrays for millimeter-wave systems that cover 24.75-27.5 GHz, as well as a dual-band 8-antenna array for sub-6 GHz systems that cover 3.3-3.6 GHz and 4.8-5.0 GHz. The frequency bands listed above will be used in future 5G communications. The simulation results suggest that both systems can function well. Antenna prototype for sub-6GHz system is built and measured, with very good agreement between simulated and measured results.

[143] Sumer.Singh Singhwal et al., introduces a dual-band circularly polarised MIMO dielectric resonator antenna (DRA) for 3.5 and 5.5 GHz bands, both of which are less than 6 GHz. They are known as sub-6 (or mid-band) and provide adequate coverage and capacity in the newly intended fifth-generation (5G) systems. Two ring DRAs (RDRAs) are etched on a 0.8 mm thick RT Duroid substrate to form the suggested structure. Broadside impedance bandwidths are 3.1-3.75 GHz (19%) and 5.3-5.6 GHz (9.4%), respectively, whereas circular polarisation (CP) bandwidths are 3.425-3.6 GHz

(5%), and 5.45-5.55 GHz (2%). CP is produced by stimulating HE modes with two probes arranged orthogonally, that is, at a 90-degree azimuthal angular distance apart. Changing the length of the probe allows for the required time-phase quadrature between modes.

[144] Hsiang Nerng Chen et al., describes an upgraded diversity wideband circularly polarised (CP) multiple-input multiple-output (MIMO) dielectric resonator antenna (DRA). Two diagonal edges of the DR were trimmed at 45° in the DRA element to achieve a broader axial ratio greater than 0.65 GHz. To generate CP fields, a cross-ring slot with a specified slot-arm ratio was excited by a microstrip-line (MSTL) implemented on the backside of the FR4 substrate. Small triangular stands were used to hold the DR in place to prevent degradation from the uncontrolled bonding chemical used to adhere the DR to the FR4 substrate. With an antenna gain of 4.83 dBi, the DRA achieved an impedance bandwidth greater than 0.8 GHz. Using the DRA and the MSTL feed, a two-element CP-DRA array with EBG structure engraved onto the ground plane of the MSTL was created. The suggested architecture offers isolation greater than 26 dB over the specified frequency band without sacrificing performance while remaining compact in the array. On the implemented circularly polarised MIMO DRA, various diversity analyses were performed. The measured findings showed that the suggested singly fed DRA with EBG on the ground plate is suitable for building wideband circular polarised MIMO antennas in a small package.

[145] Matthew Bridges et al., presents an 8x8 Multiple Input Multiple Output (MIMO) antenna design for Fifth Generation (5G) sub-6 GHz smartphone applications. The antenna elements are based on a quarter wavelength folded monopole that operates between 3.4 and 3.8 GHz. Physical separation between antenna elements provides isolation. The outside shell of the constructed antenna prototype is built of Rogers R04003C and has measurements based on future 5G equipped phones. The measured operating bandwidth is 3.32 to 3.925 GHz with a transmission coefficient of -14.7 dB. An antenna array with a high total efficiency is also produced. The architecture is appropriate for MIMO communications, as evidenced by an Envelope Correlation Coefficient (ECC) of 0.014.

[146] Jiu-Kun Che et al., highlights with a total volume of only 152 mm in diameter and 20 mm in height, proposes a compact vehicle antenna design that supports 4-channel MIMO operations for 5G Sub 6 GHz, LTE, WLAN, and V2X communications. Four dielectric-loaded broad horns, a common top plate, an annular slot, and four shorting pins comprise the proposed antenna system. In the selective combining mode, it generates four vertically polarised diversity patterns in azimuth and a relatively uniform azimuthal coverage. The suggested antenna's working principles and important design parameters are thoroughly described.

[147] R Krishnamoorthy et al., proposes a four-element MIMO antenna that operates in three Sub-6 GHz 5G channels. The single element antenna is made up of a square-shaped patch with circular and rectangular-shaped slot cuts. The single piece is symmetrically positioned along the horizontal axis and mirrored along the vertical and diagonal axes. The four-element MIMO antenna has a symmetric structure on top and a linked ground at the back. The proposed bands achieve satisfactory inter-element isolation with a directed radiation pattern and an average gain of 2.5 dBi. Furthermore, the antenna passes the MIMO diversity criteria, making it commercially viable for usage in Sub-6 GHz wireless 5G devices.

[148] Deepti Sharma et al., describes a compact six-element MIMO antenna. The proposed antenna's resonating elements are simple monopoles stimulated by microstrip feed lines. Two parts of the MIMO antenna span the frequency bands 727-1066 MHz and 5.28-6.2 GHz for IoT and WLAN/Wi-Fi applications, respectively. The other two elements include the NB-IoT band at 1.7-1.9 GHz, Wi-Fi band at 5.5-5.8 GHz, and X-band spectrum at 7.2-8.9 GHz. The other two parts span the 3.1-3.58 GHz frequency range for sub-6 GHz applications. In the proposed MIMO antenna, a decoupling network and stubs are used for isolation and impedance matching.

[149] Sanjukta Nej et al., highlights a unique microstrip line-fed meander-line-based four-elements quad band Multiple Input and Multiple Output (MIMO) antenna with a gain enhancement technique. The proposed structure resonates at four bands at the same time, namely 1.23, 2.45, 3.5, and 4.9 GHz, which correspond to GPS L2, Wi-Fi, Wi-MAX, and WLAN wireless application bands. The unit element is expanded to a four-element

MIMO antenna construction that exhibits isolation of more than 22 dB between adjacent elements while without disrupting the resonant frequencies. Two orthogonal microstrip lines are added between the antenna elements to improve gain, resulting in significant gain improvement across all four resonances. Furthermore, the MIMO structure's diversity performance is examined.

[150] Mousa Hussein et al., investigates a ten-element MIMO antenna with better isolation and X-polar radiation reduction in the millimetre area. Each antenna element is made up of a stub loaded coplanar waveguide (CPW) structure with an elliptical shaped radiator. To lessen the mutual coupling impact between the MIMO's nearby antenna pairs, a Defected Ground Structure (DGS) was integrated by etching partial elliptical shaped slots at the coplanar ground. The backside of the substrate has a metallic plane with flaws that is not connected to the antenna's coplanar ground. This defective metal structure (DMS) serves as a reflector, increasing radiation properties by lowering XPR and improving co-pol to X-pol isolation. When compared to previous mm-wave MIMO antennas, the suggested structure is found to be beneficial in terms of antenna element count, simplicity, low mutual coupling, co-pol to X-pol isolation, and ECC. The planned 10-elements MIMO prototype is built, and the measured results match the simulated results.

References

[1]. Qasim H. Kareem, Malik J. Farhan and Ali K. Jassim, "Design and Analysis of a Frequency Reconfigurable Octagonal Ring-Shaped Quad-Port Dual-Band Antenna Based on a Varactor Diode", *Progress In Electromagnetics Research C*, Vol. 116, 235–248, **2021**.

[2]. Amjad Iqbal, Amor Smida, Nazih Khaddaj Mallat, Ridha Ghayoula, Issa Elfergani, Jonathan Rodriguez and Sunghwan Kim," Frequency and Pattern Reconfigurable Antenna for Emerging Wireless Communication Systems", *Electronics*, Vol. 8, 407, **2019**; doi:10.3390/electronics8040407.

[3]. Mohammad M. Fakharian, Pejman Rezaei and Ali A. Orouji," Polarization and Radiation Pattern Reconfigurability of a Planar Monopole-Fed Loop Antenna for GPS Application", *Radio engineering*, Vol. 25, No. 4, 680-686, December **2016**; doi: 10.13164/re.2016.0680.

[4]. Mustafa Murat Bilgiç and Korkut YeLin," Polarization Reconfigurable Patch Antenna for Wireless Sensor Network Applications", *International Journal of Distributed Sensor Networks*, Article ID: 967329, Pg. No. 1-5, **2013**; doi: 10.1155/2013/967329.

[5]. Sunghwan Kim, Amjad Iqbal, Amor Smida, Lway Faisal Abdulrazak, Omar A. Saraereh, Nazih Khaddaj Mallat and Issa Elfergani," Low-Profile Frequency Reconfigurable Antenna for Heterogeneous Wireless Systems", *Electronics*, 8, 976, **2019**; doi:10.3390/electronics8090976.

[6]. Phalguni Mathur, Athira Raveendran and Sujith Raman," Mechanically Frequency Reconfigurable Antenna and its Application as a Fluid Level Detector for Wireless Sensor Networks", *URSI AP-RASC 2019*, New Delhi, India, 09 - 15 March **2019**.

[7]. Taehee Jang, Cheng Zhang, Hongseok Youn, Jing Zhou, and L. Jay Guo," Semi-transparent and Flexible Mechanically Reconfigurable Electrically Small Antennas Based on Tortuous Metallic Micromesh", *IEEE Transactions on Antennas and Propagation*, **2016**; doi: 10.1109/TAP.2016.2623479.

[8]. I.F. da Costa and D. H. Spadoti," Optically controlled reconfigurable antenna for 5G future broadband cellular communication networks", *Journal of Microwaves, Optoelectronics and Electromagnetic Applications*, Vol. 16, No. 1, March **2017**; doi:10.1590/2179-10742017v16i1883.

[9]. A.M. Mabrouk, Ahmed A. Ibrahim, Hesham F.A. Hamed, "Reconfigurable antenna with frequency and beam switching using transformer oil and PIN-diode for microwave applications", *Alexandria Engineering Journal*, Volume 61, Issue 3, Pg. No. 1824-1833, March **2022**; doi: 10.1016/j.aej.2021.06.099.

[10]. Sarawuth Chaimool, Tanan Hongnara, Chawalit Rakluea, Prayoot Akkaraekthalin,and Yan Zhao," Design of a PIN Diode-Based Reconfigurable Metasurface Antenna for Beam Switching Applications", *International Journal of Antennas and Propagation*, Article ID: 7216324, Pg.No: 1-7, **2019**; doi: 10.1155/ 2019/7216324.

[11]. Ming-Chun Tang, Zheng Wen, Hao Wang, Mei Li, Richard W. Ziolkowski," Compact, Frequency-Reconfigurable Filtenna with Sharply Defined Wideband and Continuously Tunable Narrowband States", *IEEE Transactions on Antennas and Propagation*, Vol. 65, Pg.No: 10, October **2017**; doi: 10.1109/TAP.2017.2736535.

[12]. Mahmoud Al Ahmad, Saifudeen Kabeer, Ala Abu Sanad and Lillian J. A. Olule," Compact single-varactor diode frequency-reconfigurable microstrip patch antenna", *IET Microwaves, Antennas & Propagation*, 1 March **2021**; doi: 10.1049/ mia2.12117.

[13]. Hesham A. Mohamed, Ahmed S. Elkorany, Sammar A. Saad and Demyana A. Saleeb," New Simple Flower Shaped Reconfigurable Band-Notched UWB Antenna Using Single Varactor Diode", *Progress In Electromagnetics Research C*, Vol. 76, 197-206, **2017**; doi: 10.2528/PIERC17061507.

[14]. Preet Kaur, Asok De and S. K. Aggarwal," Design of A Novel Reconfigurable Fractal Antenna for Multi-Band Application", *International Journal of Advanced Science and Technology*, Vol.62, pp.103-112, **2014**; doi: 10.14257/ ijast. 2014. 62.08.

[15]. Phalguni Mathur, Gopikrishna Madanan and Sujith Raman," Mechanically frequency reconfigurable antenna for WSN, WLAN, and LTE 2500 based internet of things applications", *Int J RF Microw Comput Aided Eng*, **2020**; doi: 10.1002/ mmce.22318.

[16]. Sonia Sharma and Chandra C. Tripathi," Frequency Reconfigurable U-slot Antenna for SDR Application", *Progress In Electromagnetics Research Letters*, Vol. 55, 129-136, **2015.**

[17]. Ting-Yan Liu, Jun-Yan Chen, and Jeen-Sheen Row," Frequency Reconfigurable Antenna with Dual-Band and Dual-Mode Operation", *Progress In Electromagnetics Research Letters*, Vol. 97, 115–120, **2021**.

[18]. Mubashir Ul Hassan, Farzana Arshad, Syeda Iffat Naqvi, Yasar Amin, Hannu Tenhunen," A Compact Flexible and Frequency Reconfigurable Antenna for Quintuple Applications", *Radio Engineering*, Vol. 26, No. 3, September **2017**; doi: 10.13164/re.2017.0655.

[19]. Guiping Jin, Chuhong Deng, Ju Yang, Yechun Xu and Shaowei Liao," A New Differentially-Fed Frequency Reconfigurable Antenna for WLAN and Sub-6 GHz 5G Applications", *IEEE Access*, Vol.7, **2019**; doi: 10.1109/ ACCESS. 2019. 29017 60.

[20]. Geriki Polaiah, Krishnamoorthy Kandasamy and Muralidhar Kulkarni," An Autonomous Frequency Reconfigurable Antenna Using Slot line Open-Loop Resonators", *IEEE Access*, Vol.4, **2016**; doi: 10.1109/ACCESS.2021.3087123.

[21]. Abu Hena Murshed, Md. Azad Hossain, Eisuke Nishiyama, Ichihiko Toyoda," Design and characterization of frequency reconfigurable honey bee antenna for cognitive radio application", *International Journal of Electrical and Computer Engineering (IJECE)*, Vol. 12, No. 6, pp. 6178-6186 ISSN: 2088-8708, December **2022**; doi: 10.11591/ijece. v12i6.pp6178-6186.

[22]. Subhash B K, R Jyosthna, R Akshay Sunny, Sunil D M, Rajasekhar C.Biradhar and Tanweer Ali," A Compact Quad-Band Frequency Reconfigurable Antenna for 5G and Satellite Applications", *International Conference on Advances in Electronics, Computers and Communications (ICAECC)*, May 14, **2021**.

[23]. Malathi Kanagasabai, Padmathilagam Sambandam, Gulam Nabi Alsath Mohammed, N. M. Dinesh, Michael Shaun Morais, Akshara Viswanathan, Sandeep Kumar Palaniswamy and Arun kumar Shrivastav," On the design of frequency reconfigurable tri-band miniaturized antenna for WBAN applications", *Int. J. Electron. Commun. (AEÜ)*, Vol.127, 153450, **2020**; doi: 10.1016/j.aeue. 2020.153450.

[24]. Dildar H, Althobiani F, Ahmad I, Khan, W.U.R, Ullah S, Mufti N, Ullah S, Muhammad F, Irfan M, Glowacz A," Design and Experimental Analysis of Multiband Frequency Reconfigurable Antenna for 5G and Sub-6 GHz Wireless Communication", *Micromachines*,12,32, **2021**; doi: 10.3390/ mi12010032.

[25]. Rahul Kumar Verma, Anubhav Kumar and R. L. Yadava," Compact Multiband CPW Fed Sub 6 GHz Frequency Reconfigurable Antenna for 5G and Specific UWB Applications", *Journal of Communications*, Vol. 15, No. 4, 345-349, April **2020**; doi:10.12720/jcm.15.4.

[26]. Poonam Thanki and Falguni Raval," Fork-shaped frequency and pattern reconfigurable antenna", *Int J Commun Syst.*, **2020**; doi: 10.1002/dac.4613.

[27]. M. Faizal Ismail, M. Kamal A. Rahim, M. Rijal Hamid, Huda A. Majid, A. Halim Omar, Levy O. Nur, Bambang S. Nugroho," Dual-Band Pattern Reconfigurable Antenna using Electromagnetic Band-Gap Structure", *International Journal of Electronics and Communications*, **2020**; doi: 10.1016/j.aeue. 2020.153571.

[28]. Jerzy Kowalewski, Jude Atuegwu, Jonathan Mayer, Tobias Mahler and Thomas Zwick," A Low-Profile Pattern Reconfigurable Antenna System for Automotive MIMO Applications", *Progress In Electromagnetics Research,* Vol. 161, 41–55, **2018**.

[29]. Jinzhi Zhou, Ming Yang and Junnan Yu," Pattern Reconfigurable Patch Antenna with Dual Band Characteristic for WLAN & 5G Applications", *Progress In Electromagnetics Research M*, Vol. 98, 147–158, **2020**.

[30]. Kushmanda Saurav, Debdeep Sarkar and Kumar Vaibhav Srivastava," A dual-band reconfigurable Yagi–Uda antenna with diverse radiation patterns", *Appl. Phys. A*, 123:483, **2017**; doi: 10.1007/s00339-017-1087-y.

[31]. Adnan Ghaffar, Xue Jun Li, Wahaj Abbas Awan, Aqeel Hussain Naqvi, Niamat Hussain, Mohammad Alibakhshikenari and Ernesto Limiti," A Flexible and Pattern Reconfigurable Antenna with Small Dimensions and Simple Layout for Wireless Communication Systems Operating Over 1.65–2.51 GHz", *Electronics*, 10, 601, **2021**; doi: 10.3390/ electronics10050601.

[32]. Kranti Kumar Katare, Animesh Biswas and Karu P. Esselle," Directive Array Based Pattern Reconfigurable Antenna", *11th European Conference on Antennas and Propagation (EUCAP)*, March **2017**, doi: 10.23919/EuCAP.2017.7928038.

[33]. Ke Li, Yan Shi and Quan Wei Wu," A pattern-reconfigurable antenna design for body area networks communications", *Int J RF Microw Comput Aided Eng.*, **2021**; doi: 10.1002/mmce.22820.

[34]. Saeed A. Haydhah, Fabien Ferrero, Leonardo Lizzi, Mohammad S. Sharawi and Azzedine Zerguine," A Multifunctional Compact Pattern Reconfigurable Antenna with Four Radiation Patterns for Sub-GHz IoT Applications", *IEEE Journal of Antennas and Propagation*, 10 May **2021**; doi: 10.1109/OJAP.2021.3078236.

[35]. Abdellah Attalhaoui, Abdelhak Bendali, Abid-Reda El Wardi and Mohamed Habibi," Dual Linear Polarization Reconfigurable Patch Antennas for Lower 5G Frequency Band", *Journal of Theoretical and Applied Information Technology*, Vol.99. No 20, SSN: 1992-8645, 31st October **2021**.

[36]. Xu Chang, Wang Zhengpeng, Wang Yi, Wang Peng and Gao Steven, "A polarization-reconfigurable wideband high-gain antenna using liquid metal tuning", *IEEE Transactions on Antennas and Propagation*, Vol. 68, No. 8, Pg.No: 5835-5841, **2020**; doi: 10.1109/TAP.2020.2996773.

[37]. Al-Yasir Yasir I.A, Ojaroudi Parchin Naser, Elfergani Issa T, Abd-Alhameed Raed A, Noras James M, Rodriguez Jonathan, Al-jzari A and Hammed W.I," A New Polarization-Reconfigurable Antenna for 5G Wireless Communications", *9th International Conference on Broadband Communications, Networks, and Systems*, Sept 19-20, **2018**; Faro, Portugal.

[38]. Zhilong Li, Yajie Mu, Jiaqi Han, Xiaohe Gao and Long Li," Dual-polarized antenna design integrated with metasurface and partially reflective surface for 5G communication", *EPJ Appl. Metamat.* 7, 3,**2020**.

[39]. Emad Al Abbas, Nghia Nguyen-Trong, Ahmed Toaha Mobashsher and Amin M. Abbosh," Polarization-Reconfigurable Antenna Array for Millimeter-Wave 5G", *IEEE Access*, September 6, **2019**; doi: 10.1109/ACCESS.2019.2939815.

[40]. Naser Ojaroudi Parchin, Yasir I. A. Al-Yasir, Abdulkareem S. Abdullah, Raed A. Abd-Alhameed and James M. Noras," A New Polarization-Reconfigurable Antenna for 5G Applications", *Electronics*,7, 293, **2018**; doi:10.3390/ electronics 7110293.

[41]. Swetha Ravikanti and L. Anjaneyulu," A Novel and Compact Circularly Polarized Antenna for 5G Wireless Local Area Network Application", *Electrical, Control and Communication Engineering*, ISSN 2255-9159, Vol. 16, No. 1, Pg. No: 44–50, **2020**; doi: 10.2478/ecce-2020-0007.

[42]. Xue Ren, Shaowei Liao and Quan Xue," A Differentially-Fed Dual-Polarized Magnetic Dipole Antenna for Spaceborne Applications", *IEEE Transactions on Antennas and Propagation*, **2018**; doi: 10.1109/TAP.2018.2883526.

[43]. He Huang, Xiaoping Li and Yanming Liu," A Low Profile, Single-ended and Dual-polarized Patch Antenna for 5G Application", *IEEE Transactions on Antennas and Propagation*, **2019**; doi: 10.1109/TAP.2019.2948743.

[44]. Khaled Yahya Alqurashi, James R. Kelly, Zhengpeng Wang, Carol Crean, Raj Mittra, Mohsen Khalily and Yue Gao, "Liquid Metal Bandwidth Reconfigurable Antenna", *IEEE Transactions on Antennas and Propagation Letters*, **2019**; doi: 10.1109/LAWP.2019.2959879.

[45]. Mohamed Bikrat and Seddik Bri," A Bandwidth Reconfigurable Planar Antenna for UWB Applications", *E3S Web of Conferences, ICIES'22*, 351, 01060, **2022**; doi: 10.1051/e3sconf/202235101060.

[46]. T. Anilkumar, B.T.P. Madhav, M. Venkateswara Rao and B. Prudhvi Nadh," Bandwidth reconfigurable antenna on a liquid crystal polymer substrate for automotive communication applications", *Int. J. Electron. Commun. (AEÜ)*, 117, 153096, **2020**; doi: 10.1016/j.aeue.2020.153096.

[47]. Ulomi George and Fang Lili, "A simple frequency and polarization reconfigurable antenna", *Electromagnetics*, 40:6, 435-444, **2020**; doi: 10.1080/ 02726343. 2020. 1811940.

[48]. Jun Hu and Zhang-Cheng Hao," Design of a Frequency and Polarization Reconfigurable Patch Antenna with a Stable Gain", Special section on Multi-function RF Components for Current and Future 5G Wireless Communications, *IEEE Access*, December 3,**2018**; doi: 10.1109/ ACCESS. 2018. 2879498.

[49]. Suresh Kumar M and Yogesh Kumar Choukiker," Frequency and Polarization Reconfigurable Microstrip Antenna with Switching Feed Configuration", *IEEE Access*, 978-1-5386-7070-6/18, **2018**.

[50]. Singh G, Kanaujia B K, Pandey V K, Gangwar D, Kumar S," Pattern and frequency reconfigurable antenna with diode loaded ELC resonator", *International Journal of Microwave and Wireless Technologies*, 1–13, **2019**; doi: 10.1017/S175 90787190 01077.

[51]. MLSNS Lakshmi, B T P Madhav, Habibulla Khan and P V V Kishore," A Frequency and Pattern Reconfigurable Asymmetric Ground Antenna on Flexible Polyimide Material for LTE, Wi-Fi, WLAN and Fixed Satellite Applications", *Flexible and Printed Electronics*, KFPE, **2020**; doi: 10.1088/2058-8585/ab8f35.

[52]. Anuradha A. Palsokar and S.L. Lahudkar, "Frequency and pattern reconfigurable rectangular patch antenna using single PIN diode", *Int. J. Electron. Commun. (AEÜ)*, 125, 153370, **2020**; doi: 10.1016/j.aeue.2020.153370.

[53]. Yixin Li, Yu-Lin Chang and Qing-Xin Chu," Reconfigurable dipole antenna array with shared parasitic elements for 360-degree beam steering", *Int J RF Microw Comput Aided Eng.*, 22 August **2020**; doi: 10.1002/mmce.22435.

[54]. Son Xuat Ta, Kam Eucharist Kedze, Dao Ngoc Chien and Ikmo Park," Bandwidth-Enhanced Low-Profile Antenna with Parasitic Patches", *International Journal of Antennas and Propagation*, Article ID: 6529060, 11 pages, **2017**; doi: 10.1155/ 2017/6529060.

[55]. Youcef Braham Chaouche, Idris Messaoudene, Ismail Benmabrouk, Mourad Nedil and Farid Bouttout," Compact coplanar Waveguide-fed reconfigurable fractal antenna for switchable multiband systems", *IET Microwaves, Antennas and Propagation*, **2018**; doi: 10.1049/iet-map.2018.5005.

[56]. Harish Chandra Mohanta, Abbas Z. Kouzani and Sushanta K. Mandal," Reconfigurable Antennas and Their Applications", *Universal Journal of Electrical and Electronic Engineering,* 6(4): 239-258, **2019**; doi: 10.13189/ ujeee. 2019. 060406.

[57]. Bouyedda A, Barelaud B and Gineste L," Design and Realization of an UHF Frequency Reconfigurable Antenna for Hybrid Connectivity LPWAN and LEO Satellite Networks", *Sensors*, 21, 5466, **2021**; doi: 10.3390/s21165466.

[58]. Jeong Keun Ji," Dual-band pattern reconfigurable antenna for wireless MIMO applications", *The Korean Institute of Communications Information Sciences*, Volume 2, Issue 4, Pg. No. 199-203, December **2016**; doi: 10.1016/ j.icte. 2016. 08. 002.

[59]. Muhammad Ikram, Emad Al Abbas, Nghia Nguyen-Trong, Khalil H. Sayidmarie and Amin Abbosh, "Integrated Frequency-Reconfigurable Slot Antenna and Connected Slot Antenna Array for 4G and 5G Mobile Handsets", *IEEE Transactions on Antennas and Propagation*, **2019**; doi: 10.1109/ TAP. 2019. 2930 119.

[60]. Qasim Awais, Hassan Tariq Chattha, Mohsin Jamil, Yang Jin, Farooq Ahmad Tahir and Masood Ur Rehman, "A Novel Dual Ultrawideband CPW-Fed Printed Antenna for Internet of Things (IoT) Applications", *Hindawi, Wireless Communications and Mobile Computing*, Article ID: 2179571, 9 pages, **2018**; doi: 10.1155/ 2018/ 2179571.

[61]. Arjuna Muduli, Murali Krishna Kanneboina, Krishna Vamsi Mudumunthala and Sriram Valluri," A Reconfigurable Wearable Antenna for Mid Band 5G Applications", *Journal of Physics: Conference Series, IOP Publishing,* 1921, 012051, **2021**; doi:10.1088/1742-6596/1921/1/012051.

[62]. Mingyu Sun, Zhe Zhang, Fan Zhang and Aixin Chen," L/S Multiband Frequency-Reconfigurable Antenna for Satellite Applications", *IEEE Antennas and Wireless Propagation Letters*, **2019**; doi: 10.1109/LAWP.2019.2945624.

[63]. Muhammad Yousaf, Naeem Abbas, Farooq Ali, Muhammad Munwar Iqbal, Yasar Amin and Adeel Akram," Miniaturized High Gain Antenna for Biomedical Application", *ICFNDS'19,* Paris, France, July 1-2, **2019**.

[64]. Hassan T. Chattha," Compact high isolation wideband 4G and 5G multi-input multi-output antenna system for handheld and internet of things applications*", Int J RF Microw Comput Aided Eng.,* 15 March **2019**; doi: 10.1002/mmce.21710.

[65]. Narinder Sharma and Sumeet Singh Bhatia," Performance enhancement of nested hexagonal ring-shaped compact multiband integrated wideband fractal antennas for wireless applications", *Int J RF Microw Comput Aided Eng.*, **2019**; doi: 10.1002/ mmce.22079.

[66]. Sreenath Reddy Thummaluru, Mohammad Ameen and Raghvendra Kumar Chaudhary," Four-Port MIMO Cognitive Radio System for Mid-Band 5G Applications", *IEEE Transactions on Antennas and Propagation*, **2019**; doi: 10.1109/TAP.2019.2918476.

[67]. Nand Kishore, Arun Prakash and V. S. Tripathi," A Multiband Microstrip Patch antenna with Defected Ground Structure for its Applications", *Microwave and Optical Technology Letters*, Vol. 58, No. 12, December **2016**; doi: 10.1002/ mop. 30151.

[68]. Eunyoung Park and Sangkil Kim," Design and Analysis of a TEM Mode Rectangular Coaxial Waveguide for Mobile 5G Millimeter Wave Antenna Module Applications", *Journal of Electromagnetic Engineering and Science*, Vol. 20, No. 3, 169-175, July **2020**; doi: 10.26866/jees.2020.20.3.169.

[69]. He Huang, Xiaoping Li and Yanming Liu," A Low Profile, Single-ended and Dual-polarized Patch Antenna for 5G Application" *IEEE Transactions on Antennas and Propagation*, **2019**; doi: 10.1109/TAP.2019.2948743.

[70]. Jose MC, Sankararajan R, Sreeja BS, Alsath MGN, Kumar P, "A compact omnidirectional to directional frequency reconfigurable antenna for wireless sensor network applications", *International Journal of Microwave and Wireless Technologies*, 1–12, **2021**; doi: 10.1017/ S1759078721000994.

[71]. Aabia Khan, Maryam Rasool, Farooq Bhatti, Bilal Ijaz and Adnan Iftikhar," A Compact Circular Loop Inspired Frequency and Bandwidth Reconfigurable Antenna for 4G, 5G, and X- Band Applications", *Radio Engineering*, Vol. 29, No. 3, September **2020**; doi: 10.13164/re.2020.0471.

[72]. Kumar S, Lee GH, Kim DH, Mohyuddin W, Choi HC and Kim KW," A compact four-port UWB MIMO antenna with connected ground and wide axial ratio bandwidth", *International Journal of Microwave and Wireless Technologies*, 1–11, **2019**; doi: 10.1017/S1759078719000874.

[73]. Niraj Kumar and Usha Kiran Kommuri," MIMO Antenna H-Plane Isolation Enhancement using UC-EBG Structure and Metal Line Strip for WLAN Applications", *Radio Engineering*, Vol. 28, No. 2, June **2019**; doi: 10.13164/re. 2019.0399.

[74]. Charles MacWright Thomas, Huda A. Majid, Zuhairiah Zainal Abidin, Samsul Haimi Dahlan, Mohamad Kamal A. Rahim and Raimi Dewan," A Study on V-Shaped Microstrip Patch MIMO Antenna", *Indonesian Journal of Electrical Engineering and Computer Science,* Vol. 5, No. 3, pp. 606 – 611, March **2017**; doi: 10.11591/ijeecs.v5.i3.pp606-611.

[75]. Alaa M. Hediya, Ahmed M. Attiya and Walid S. El-Deeb," Multiple-Input Multiple-Output Antenna for Sub-Six GHz 5G Applications Using Coupled Folded Antenna with Defective Ground Surface", *Progress In Electromagnetics Research C*, Vol. 114, 13–29, **2021**.

[76]. Leeladhar Malviya, Rajib K. Panigrahi, and Machavaram V. Kartikeyan," A 2 × 2 Dual-Band MIMO Antenna with Polarization Diversity for Wireless Applications", *Progress In Electromagnetics Research C*, Vol. 61, 91–103, **2016**.

[77]. Muhammad Jaafer Riaz, Ayesha Sultan, Muhammad Zahid, Anum Javed, Yasar Amin and Jonathan Loo," MIMO Antennas for Future 5G Communications", *IEEE 23rd International Multitopic Conference*, **2020**; doi: 10.1109/ INMIC 50486.2 020. 9318126.

[78]. Amjad Iqbal, Ahsan Altaf, Mujeeb Abdulla, Mohammad Alibakhshikenari, Ernesto Limiti and Sunghwan Kim," Modified U-Shaped Resonator as Decoupling Structure in MIMO Antenna", *Electronics*, 9, 1321, **2020**; doi:10.3390/ electronics 9081321.

[79]. Huey Shin Wong, Mohammad Tariqul Islam, and Salehin Kibria," Design and Optimization of LTE 1800 MIMO Antenna", *The Scientific World Journal Volume*, Article ID: 725806, **2014**; doi: 10.1155/2014/725806.

[80]. Xiongwen Zhao and Sharjeel Riaz," A Dual-Band Frequency Reconfigurable MIMO Patch-Slot Antenna Based on Reconfigurable Microstrip Feedline", *IEEE Access*, **2018**; doi: 10.1109/ACCESS.2018.2858442.

[81]. Aftab Ahmad Khan, Mohd Haizal Jamaluddin, Sajid Aqeel, Jamal Nasir, Jalil ur Rehman Kazim and Owais Owais," Dual band MIMO dielectric resonator antenna for WiMAX/WLAN applications, *IET Microwaves Antennas and Propagation*, January **2017**; doi: 10.1049/iet-map.2015.0745.

[82]. Yixin Li, Chow-Yen-Desmond Sim, Yong Luo and Guangli Yang," 12-Port 5G Massive MIMO Antenna Array in Sub-6GHz Mobile Handset for LTE Bands 42/43/46 Applications", *IEEE Access*, **2018**; doi: 10.1109/ ACCESS. 2017.2763161.

[83]. Gourab Das and Ravi Kumar Gangwar," A High Isolation MIMO Cylindrical Dielectric Resonator Antenna for 4G Applications", *3rd International Conference on Microwave and Photonics (ICMAP 2018)*, 9-11 February, **2018**, doi: 10.1109/ ICMAP.2018.8354511.

[84]. Jie-Huang Huang, Wen-Jiun Chang and Christina F. Jou," Dual-Band MIMO Antenna with High Isolation Application by Using Neutralizing Line", *Progress In Electromagnetics Research Letters*, Vol. 48, Pg.No 15–19, **2014**.

[85]. Irene Kong Cheh Lin, Mohd Haizal Jamaluddin, Azlan Awang, Raghuraman Selvaraj, Muhammad Hashim Dahri, Leow Chee Yen and Hasliza A. Rahim," A Triple Band Hybrid MIMO Rectangular Dielectric Resonator Antenna for LTE Applications", *IEEE Access*, Vol.4, **2016**; doi: 10.1109/ACCESS.2019.2937987.

[86]. Adham M. Saleh, Tariq A. Nagem, Raed A. Abd-Alhameed, James M. Noras and Chan H. See," Mutual Coupling Reduction of Dual-Band Uni-Planar MIMO System Using Neutralization Line Technique", *Radio Communication*, October **2020**.

[87]. Tripta Kumari, Gourab Das, Ravi K. Gangwar and Kundan K. Suman," Dielectric resonator based two-port dual band antenna for MIMO applications", *Int J RF Microw Comput Aided Eng*., December **2019**; doi: 10.1002/mmce.21985.

[88]. Ming-Yang Li, Yong-Ling Ban, Zi-Qiang Zu, Gang Wu, Chow-Yen-Desmond Sim, Kai Kang and Zhe-Feng Yu, "Eight-Port Orthogonally Dual-Polarized Antenna Array for 5G Smartphone Applications," in *IEEE Transactions on Antennas and Propagation*, Vol. 64, No. 9, Pg.No: 3820-3830, September **2016**; doi: 10.1109/TAP.2016.2583501.

[89]. Iqbal A, Nasir J, Qureshi MB, Khan AA, Rehman JU and Owais," A CPW fed quadport MIMO DRA for sub-6 GHz 5G applications", *PLoS ONE* 17(6): e0268867, **2022**; doi: 10.1371/journal.pone.0268867.

[90]. Leeladhar Malviya, Ajay Parmar, Deepak Solanki, Parul Gupta and Priyanshi Malviya," Highly isolated inset-feed 28 GHz MIMO-antenna array for 5G wireless application", *Procedia Computer Science*, 171, 1286–1292, **2020**.

[91]. M V Karthikeyan, L. Malviya, R.K. Panigrahi, "*Multi-standard, multi-band planar multiple input multiple output antenna with diversity effects for wireless applications," Int. J. RF Microwave Computer Aided Engineering*, September **2018**; doi: 10.1002/mmce.21551.

[92]. Nikesh Kumar Sahu, Gourab Das and Ravi Kumar Gangwar," Dual polarized triple-band dielectric resonator-based hybrid MIMO antenna for WLAN/WiMAX applications", *Microwave and Optical Technology Letters*, April **2018**; doi: 10.1002/mop.31102.

[93]. Su Xu, Ming Zhang, Huailin Wen and Jun Wang," Deep-subwavelength Decoupling for MIMO Antennas in Mobile Handsets with Singular Medium", *Nature Scientific Reports*, 7: 12162, 22 September **2017**; doi:10.1038/s41598-017-11281-2.

[94]. Naser Ojaroudi Parchin, Haleh Jahanbakhsh Basherlou, Yasir I. A. Al-Yasir, Atta Ulla, Raed A. Abd-Alhameed and James M. Noras, "Multi-Band MIMO Antenna Design with User-Impact Investigation for 4G and 5G Mobile Terminals", *Sensors*, 19, 456, **2019**; doi:10.3390/s19030456.

[95]. Pawan Kumar, Shabana Urooj and Fadwa Alrowais," Design of Quad-Port MIMO/Diversity Antenna with Triple-Band Elimination Characteristics for Super-Wideband Applications", *Sensors*, 20, 624,**2020**; doi:10.3390/s20030624.

[96]. Naveen Jaglan, Samir Dev Gupta, and Mohammad S. Sharawi," 18 Element Massive MIMO/Diversity 5G Smartphones Antenna Design for Sub-6 GHz LTE Bands 42/43 Applications", *IEEE Open Journal of Antennas and Propagation*, **2021**; doi: 10.1109/OJAP.2021.3074290.

[97]. Rifaqat Hussain, Ali T. Alreshaid, Symon K. Podilchak and Mohammad S. Sharawi," Compact 4G MIMO antenna integrated with a 5G array for current and future mobile handsets", *IET Microwaves, Antennas & Propagation*, ISSN 1751-8725, Vol. 11, Issue 2, Pg.No. 271–279, **2017**; doi: 10.1049/iet-map.2016.0738.

[98]. Sharjeel Riaz and Xiongwen Zhao, "An Eight-Port Frequency Reconfigurable MIMO Slot Antenna with Multi-Band Tuning Characteristics", *International Symposium on Antennas, Propagation and EM theory*, December **2018**; doi: 10.1109/ISAPE.2018.8634212.

[99]. Peng Yang, Kui.Xi Yan, Y Y Gan and Feng Yang, "Dual-band Eight-antenna MIMO array for 5G smartphone", *IEEE International Symposium on Antennas and Propagation*, July **2019**.

[100]. Haleh Jahanbakhsh Basherlou, Naser Ojaroudi Parchin and Raed A. Abd-Alhameed," MIMO Monopole Antenna Design with Improved Isolation for 5G WiFi Applications", *International Journal of Emerging Engineering Research and Technology*, Volume 7, Issue 12, Pg.No 1-5, **2019**.

[101]. Naser Ojaroudi Parchin, Haleh Jahanbakhsh Basherlou and Raed A. Abd-Alhameed," Smartphone Antenna Design Covering 2G-5G Mobile Terminals", *International Journal of Electrical and Electronic Science* ,7(x1): 1-6, ISSN: 2375-2998, **2020**.

[102]. Atta Ullah, Naser Ojaroudi Parchin, Waqas Manan, Yasir I. A. Al-Yasir and Raed A. Abd-Alhameed," F-Shape Dual-Band MIMO Antenna System for Next Generation Smartphone Applications", *2nd International Multi-disciplinary Conference Proceedings*, September **2021**.

[103]. Habib Ullah, Saeed Ur Rahman, Qunsheng Cao, Ijaz Khan and Hamid Ullah," Design of SWB MIMO Antenna with Extremely Wideband Isolation", *Electronics*, 9, 194, **2020**; doi:10.3390/electronics9010194.

[104]. Alibakhshikenari M, Virdee B. S, Shukla P, See C. H, Abd-Alhameed R and Khalily, M, "Interaction between closely packed Array antenna elements using meta-surface for applications such as MIMO systems and synthetic aperture radars", *Radio Science*, 53, 1368–1381, **2018**; doi: 10.1029/ 2018RS006533.

[105]. Haq Nawaz and Ibrahim Tekin, "Dual Port Disc Monopole Antenna for Wide - Band MIMO Based Wireless Applications", *Microwave and Optical Technology Letters*, **2017**.

[106]. Vigneswaran Dhasarathan, Truong Khang Nguyen, Manish Sharma, Shobhit K. Patel, Sudesh Kumar Mittal and M. Thurai Pandian," Design, analysis and characterization of four port multiple-input-multiple-output UWB-X band antenna with band rejection ability for wireless network applications", *Wireless Networks*, 26:4287–4302, **2020**; doi: 10.1007/s11276-020-02343-x.

[107]. Ravindra Bakale, Anil Nandgaonkar, Shankar Deosarkar and Mahesh Munde," Design of Ultra-Wideband MIMO Antenna with Dual Band Elimination Characteristics and Low Mutual Coupling", *Progress In Electromagnetics Research C*, Vol. 123, 237–251, **2022**; doi: 10.2528/PIERC22062202.

[108]. Muhammad K. Khan and Quanyuan Feng," A Novel Two Port MIMO Antenna Having Dual Stopped-Band Functionality and Enhanced Isolation", *Progress In Electromagnetics Research M*, Vol. 113, 173–185, **2022**; doi:10.2528/ PIERM22 071403.

[109]. Hrishikesh Narayan Tripathi and Kumud Ranjan Jha," Design of Compact Printed Wideband Antenna for 5.2/5.8 GHz WLAN and 3.5/5.5 GHz WiMAX Wireless USB Dongle Application", *IEEE Applied Electromagnetics Conference*, **2014**; doi: 10.1109/AEMC.2013.7045037.

[110]. Phalguni Mathur, Robin Augustine, M. Gopikrishna and Sujith Raman," Dual MIMO Antenna System for 5G Mobile Phones, 5.2 GHz WLAN, 5.5 GHz WiMAX and 5.8/6 GHz WiFi Applications", *IEEE Access*, Volume 4, **2021**; doi: 10.1109/ACCESS.2017.

[111]. Libin Sun, Yue Li, Zhijun Zhang and Hanyang Wang, "Self-Decoupled MIMO Antenna Pair with Shared Radiator for 5G Smartphones", *IEEE Transactions on Antennas and Propagation*, **2019**; doi: 10.1109/TAP.2019.2963664.

[112]. Anping Zhao, Zhouyou Ren and Shengjie Wu," Broadband MIMO Antenna System for 5G Operations in Mobile Phones", *Int J RF Microw Comput Aided Eng*, 29(10), **2019**; doi: 10.1002/mmce.21857.

[113]. Wei Hu, Long Qian, Steven Gao, Le-Hu Wen, Qi Luo, Hang Xu, Xue-Kang Liu, Ying Liu and Wei Wang," Dual-Band Eight-Element MIMO Array Using Multi-Slot Decoupling Technique for 5G Terminals", *IEEE Access*, **2017**; doi: 10.1109/ACCESS.2019.2948639.

[114]. Z. Ren and A. Zhao, "Dual-Band MIMO Antenna with Compact Self-Decoupled Antenna Pairs for 5G Mobile Applications," *IEEE Access*, Vol. 7, Pg. No: 82288-82296, **2019**; doi: 10.1109/ACCESS.2019.2923666.

[115]. Samir Salem Al-Bawri, Mohammad Tariqul Islam, Mandeep Jit Singh, Emad Alyan, Muzammil Jusoh, Thennarasan Sabapathy, Surentiran Padmanathan and Kabir Hossain," Broadband Sub-6GHz Slot-based MIMO Antenna for 5G NR Bands Mobile Applications", *Journal of Physics: Conference Series* 1962, 012038, *IOP Publishing*, **2021**; doi:10.1088/1742-6596/1962/1/012038.

[116]. Huanqing You, Yixin Li, Bin Xu, Yuanqing Chen, Hao Jin, Guangli Yang and Yong Luo, "Dual-Functional MIMO Antenna Array with High Isolation for 5G/WLAN Applications in Smartphones", *IEEE Access*, **2019**; doi: 10.1109/ ACCESS.2019.2953311.

[117]. Xuekang Liu, Wei Hu, Steven Gao, Le-Hu-Wen, Long Qian, Tianxi Feng, Rui Xu, Peng Fei and Ying Liu "Dual-Band Ten-Element MIMO Array Based on Dual-Mode IFAs for 5G Terminal Applications," *IEEE Access*, Vol. 7, Pg. No: 178476-178485, **2019**; doi: 10.1109/ACCESS.2019.2958745.

[118]. A. Ren, Y. Liu, H. -W. Yu, Y. Jia, C. -Y. -D. Sim and Y. Xu, "A High-Isolation Building Block Using Stable Current Nulls for 5G Smartphone Applications," *IEEE Access*, Vol. 7, Pg.No: 170419-170429, **2019**; doi: 10.1109/ACCESS.2019.2955495.

[119]. Raj Kumar and Gopal Surushe," Design of microstrip-fed printed UWB diversity antenna with tee crossed shaped structure", *Engineering Science and Technology, an International Journal*, **2015**; doi: 10.1016/j.jestch.2015.10.006.

[120]. S. Jaisiva, T. Rampradesh, P. Vimala, P. Pugazhendiran and SakthiSudhan Karuppanan," Epoxy based on low profile MIMO antenna for cellular systems", *Materials Today: Proceedings,* **2021**; doi: 10.1016/j.matpr.2021.02.628.

[121]. Qurrat-Ul-Ain Nadeem, Abla Kammoun, Mérouane Debbah and Mohamed-Slim Alouini," Design of 5G Full Dimension Massive MIMO Systems", *IEEE Transactions on Communications*, Vol. 66, No. 2, February, **2018**; doi: 10.1109/TCOMM.2017.2762685.

[122]. Hussain R, Sharawi M. S and Shamim A, "4-Element Concentric Pentagonal Slot-Line-Based Ultra-Wide Tuning Frequency Reconfigurable MIMO Antenna System", *IEEE Transactions on Antennas and Propagation*, *66*(8), 4282-4287, **2018**; doi: 10.1109/tap.2018.2839970.

[123]. Sharawi, M S, Hussain, R, Alreshaid, A T and Podilchak, S, "A Compact 4G MIMO Antenna Integrated with a 5G Array for Current and Future Mobile Handsets", *IET Microwaves, Antennas and Propagation*, Vol. 11, No. 2, pp. 271–279, **2017**; doi: 10.1049/iet-map.2016.0738.

[124]. Le Chang, Yafang Yu, Kunpeng Wei, and Hanyang Wang, "Polarization-Orthogonal Co-frequency Dual Antenna Pair Suitable for 5G MIMO Smartphone With Metallic Bezels", *IEEE Transactions on Antennas and Propagation*, Vol. 67, No. 8, August **2019**; doi: 10.1109/TAP.2019.2913738.

[125]. Syrytsin I, Zhang S and Pedersen G. F, "Performance Investigation of a Mobile Terminal Phased Array with User Effects at 3.5 GHz for LTE Advanced", *IEEE Antennas and Wireless Propagation Letters*, 1847 – 1850, **2017**; doi: 10.1109/ LAWP.2016.2570418.

[126]. Chenyin Yu, Shuhui Yang, Yinchao Chen, Wensong Wang, Li Zhang, Bin Li and Ling Wang," A Super-Wideband and High Isolation MIMO Antenna System Using a Windmill-Shaped Decoupling Structure", *IEEE Access*, **2019**; doi: 10. 1109/ ACCESS.2020.3004396.

[127]. Wenfei Yin, Shaoxiang Chen, Junjie Chang, Chunhua Li and Salam K. Khamas, "CPW Fed Compact UWB 4-Element MIMO Antenna with High Isolation", *Sensors*, 21, 2688, **2021**; doi: 10.3390/s21082688.

[128]. Zhongliang Deng, Chengqi Lai, Yucheng Wang and Kun Deng," Design of a Quadruple Band-Notched Ultra-Wideband Antenna Using Curled C-Shaped Structures and Interdigital Inductance Slots", *Electronics*, 11, 3949, **2022**; doi:10.3390/ electronics11233949.

[129]. Tayyab Shabbir, Mohammad Tariqul Islam, Samir Salem Al-Bawri, Rabah W. Aldhaheri, Khalid H. Alharbi, Abdulah Jeza Aljohani and Rashid Saleem," 16-Port Non-planar MIMO Antenna System with Near-Zero-Index (NZI) Metamaterial Decoupling Structure for 5G Applications", *IEEE Access*, **2020**; doi: 10.1109/ ACCESS.2020.3020282.

[130]. Reza Karimian and Hamed Tadayon, "Multiband MIMO Antenna System with Parasitic Elements for WLAN and WiMAX Application", *International Journal of Antennas and Propagation*, Article ID: 365719, **2013**; doi: 10.1155/2013/365719.

[131]. Anil Kumar Gautam, Swati Yadav and Karumudi Rambabu, "Design of Ultra-compact UWB antenna with band-notched characteristics for MIMO applications", IET Microwaves, Antennas and Propagation, **2018**; doi: 10.1049/iet-map. 2018. 0012.

[132]. Jianfeng Zhu, Botao Feng, Biao Peng, Li Deng and Shufang Li," Multiband Printed Mobile MIMO Antenna for WWAN and LTE Applications", *Microwave and Optical Technology Letters*, Vol. 59, No. 6, June **2017**; doi: 10.1002/mop.

[133]. Jun-Yi Jiang and Hsin-Lung Su," A Wideband Eight-Element MIMO Antenna Array in 5G NR n77/78/79 and WLAN-5GHz Bands for 5G Smartphone Applications", *International Journal of Antennas and Propagation*, Article ID: 8456936, **2022**; doi: 10.1155/2022/8456936.

[134]. Jayshri Kulkarni, Abdullah G. Alharbi, Arpan Desai, Chow-Yen-Desmond Sim and Ajay Poddar," Design and Analysis of Wideband Flexible Self-Isolating MIMO Antennas for Sub-6 GHz 5G and WLAN Smartphone Terminals", *Electronics*, *10*(23), 3031, **2021**; doi: 10.3390/electronics10233031.

[135]. Surentiran Padmanathan, Ahmed Mohamed Elshirkasi, Samir Salem AL-Bawri, Mohammad Tariqul Islam, Thennarasan Sabapathy, Muzammil Jusoh, Prayoot Akkaraekthalin and ping jack soh," Compact Multiband Reconfigurable MIMO Antenna for Sub- 6GHz 5G Mobile Terminal", *IEEE Access*, **2022**; doi: 10.1109/ ACCESS.2022.3180048.

[136]. Xia0-Ting Yuan, Z. Chen, T. Gu, Y. Zhao and T. Yuan, "Wideband MIMO Antenna for 5G Smartphone Applications," *2021 International Symposium on Antennas and Propagation (ISAP)*, Taiwan, pp. 1-2, **2021**; doi: 10.23919/ISAP4 7258.2021.9614401.

[137]. Ashfaq Ahmad, Dong-you Choi and Sadiq Ullah," A compact two elements MIMO antenna for 5G communication", *Nature portfolio*, 12:3608, **2022**; doi: 10.1038/ s41598-022-07579-5.

[138]. Fakharian M M, Alibakhshikenari M, See C H, Abd-Alhameed R," A high gain multiband offset MIMO antenna based on a planar log-periodic array for Ku/K-band applications", *Sci Rep*, 12(1):4044, March 8, **2022**; doi: 10.1038/s41598-022-07866-1.

[139]. Samir Salem Al-Bawri, Mohammad Tariqul Islam, Mandeep Jit Singh, Emad Alyan, Muzammil Jusoh, Thennarasan Sabapathy, Surentiran Padmanathan and Kabir Hossain, "Broadband Sub-6 GHz Slot-based MIMO Antenna for 5G NR Bands Mobile Applications", *Journal of Physics: Conference Series, IOP Publishing*, 1962, 012038, **2021**; doi: 10.1088/1742-6596/1962/1/012038.

[140]. Demos Serghiou, Mohsen Khalily, Vikrant Singh, Ali Araghi and Rahim Tafazolli," Sub-6 GHz Dual-Band 8×8 MIMO Antenna for 5G Smartphones", *IEEE Antennas and Wireless Propagation Letters*, **2020**; doi: 10.1109/ LAWP.20 20.3008962.

[141]. Kumud R. Jha, Z. A. P. Jibran, Chitra Singh and Satish K. Sharma," 4-Port MIMO Antenna using Common Radiator on a Flexible Substrate for Sub-1GHz, Sub-6 GHz 5G NR, and Wi-Fi 6 Applications", *IEEE Journal of Antennas and Propagation*, **2021**; doi: 10.1109/OJAP.2021.3083932.

[142]. Z. Ren, S. Wu and A. Zhao, "Coexist Design of Sub-6GHz and Millimeter-Wave Antennas for 5G Mobile Terminals," *2018 International Symposium on Antennas and Propagation (ISAP)*, Busan, Korea (South), pp. 1-2, **2018**.

[143]. Sumer Singh Singhwal, Binod K. Kanaujia, Ajit Singh, Jugul Kishor and Ladislau Matekovits," Dual-band circularly polarized MIMO DRA for sub-6 GHz applications", *International Journal of RF Microwave Computer Aided Engineering*, e22350, **2020**; doi: 10.1002/mmce.22350.

[144]. H. N. Chen, J. -M. Song and J. -D. Park, "A Compact Circularly Polarized MIMO Dielectric Resonator Antenna Over Electromagnetic Band-Gap Surface for 5G Applications," *IEEE Access*, Vol. 7, pp. 140889-140898, **2019**; doi: 10.1109/ ACCESS.2019.2943880.

[145]. Matthew Bridges, Mohsen Khalily, Mohammad Abedian, Demos Serghiou, Pei Xiao and Rahim Tafazolli," High Isolation 8×8 MIMO Antenna Design for 5G Sub-6 GHz Smartphone Applications", *IEEE Access*, **2020**; doi: 10.1109/ UCET51 115.2020.9205450.

[146]. Jiu-Kun Che, Chi-Chih Chen and John F. Locke," A Compact 4-Channel MIMO 5G Sub-6 GHz/ LTE/ WLAN/ V2X Antenna Design for Modern Vehicles", *IEEE Transactions on Antennas and Propagation*, **2021**; doi: 10.1109/ TAP. 2021.308 3765.

[147]. R. Krishnamoorthy, Arpan Desai, Riki Patel and Amit Grover," 4 Element compact triple band MIMO antenna for sub-6 GHz 5G wireless applications", *Wireless Networks*, 27:3747–3759, **2021**; doi: 10.1007/s11276-021-02734-8.

[148]. Deepti Sharma, Binod Kumar Kanaujia and Sachin Kumar," Compact multi-standard planar MIMO antenna for IoT/WLAN/Sub-6 GHz/X-band applications", *Wireless Networks,* **2021**; doi: 10.1007/s11276-021-02612-3.

[149]. Sanjukta Nej, Anumoy Ghosh, Sarosh Ahmad, Jayendra Kumar, Adnan Ghaffar and Mousa Hussein," Design and Characterization of 10-Elements MIMO Antenna with Improved Isolation and Radiation Characteristics for mm-Wave 5G Applications", *IEEE Access*, **2022**; doi: 10.1109/ACCESS.2022.3225446.

[150]. Mousa Hussein, Sanjukta Nej, Anumoy Ghosh, Sarosh Ahmad, Jayendra Kumar and Adnan Ghaffar, "Design and Characterization of 10-Elements MIMO Antenna with Improved Isolation and Radiation Characteristics for mm-Wave 5G Applications", *IEEE Access,* **2017**; doi: 10.1109/ACCESS.2022.3225446.

Chapter III

CHAPTER III

A COMPACT AND CLOSELY SPACED FOUR-PORT MIMO ANTENNA FOR 5G APPLICATION

A Novel high performance four-port multiple input multiple output antenna system is suggested operating in C-band for future 5G application functioning at a 3.5 GHz band. The suggested design measures an inclusive volume of 96.08 mm x 96.08 mm x 1.34 mm. The antenna is having a high impedance matching whose isolation is about -27 dB, envelope correlation coefficient (ECC) is computed and is smaller than 0.1, channel capacity loss (CCL) average value of less than 0.2 bits per second per hertz and the diversity gain (DG) of about 10 dB. The typical peak realized antenna gain of the offered MIMO antenna is also delivered with a high radiation efficiency at the frequency of 3.5 GHz. The reflection coefficient, mutual coupling, radiation pattern, current distribution and gain of antennas are all measured and explained. The design is having a compact high volume and adequate bandwidth with good accomplished gain making the antenna a very strong for 5G application.

3.1 CONCEPT OF THE DESIGNED ANTENNA

3.1.1 Design of the Antenna Geometry

The suggested antenna components are an evolution of the fundamental rectangular patch antenna. The well-known transmission line theory is used to calculate the precise dimensions of the proposed MIMO antenna [1-3]. The following theory, which is presented in equations (3.1) and (3.2), is used to compute the optimum resonant length and width for a given resonant frequency.

$$L_{re} = \frac{c}{2f_r\sqrt{\left[\frac{\varepsilon_r+1}{2}+\frac{\varepsilon_r-1}{2}\left(1+12\frac{h}{W}\right)^{-0.5}\right]}} - 2\Delta L \qquad (3.1)$$

$$W = \frac{1}{2f_r\sqrt{\mu_0\varepsilon_0}}\sqrt{\frac{2}{\varepsilon_r+1}} \qquad (3.2)$$

Similarly, h is the thickness of the substrate, c is the speed of light, f_r is the resonance frequency, and L represents the differential change in length due to fringing. Furthermore, ε_o, ε_r and μ_o are the free space permittivity, relative permittivity, and permeability of free space, respectively. The depicted antenna's dimensions for 3.5 GHz were determined using the calculations mentioned above. The design history of the four P-shaped antenna structure that make up the four element MIMO antenna is shown in Figure 3.1. The desired 3.5 GHz MIMO antenna is created on an FR-4 dielectric substrate. The proposed configuration of the four division MIMO antenna is considered as an excellent potential design in 5G antennas, with a total substrate dimension of 96.08 x 96.08 x 1.34 mm^3. The microstrip design is shown in Figure 3.1 as a specific 50 Ω microstrip line that is built and produced on a FR-4 material. Initially, only one antenna element was taken into consideration. The antenna's dimensions were tuned for 3.5 GHz at the needed frequency.

A fabricated model of the single antenna element can be seen in Figure 3.2 (a) and (b). The antenna's simulated and measured return loss are shown in Figure 3.2 (c) and (d). The simulated and measured results had a nearly identical resonance frequency, as shown in the figures. The bandwidth of the built prototype is less than that of the inferred prototypes, but it is still larger than the required band. The four P-shaped parts are symmetrically positioned with a separation distance. When this naming is employed, the least ground area is printed on the backside of the material. The ground plane of each radiating element was divided into four rectangles below it. Figure 3.1 depicts the theoretical perspective of the evaluated radiating element, which is in the front, and the partial ground, which is on the backside, as well as the ideal size. The present design is simulated using Ansys HFSS. The comparison of different 5G MIMO antennas is given in Table 3.1 and the sizes of geometrical parameters are detailed in Table 3.2 along with their dimensions.

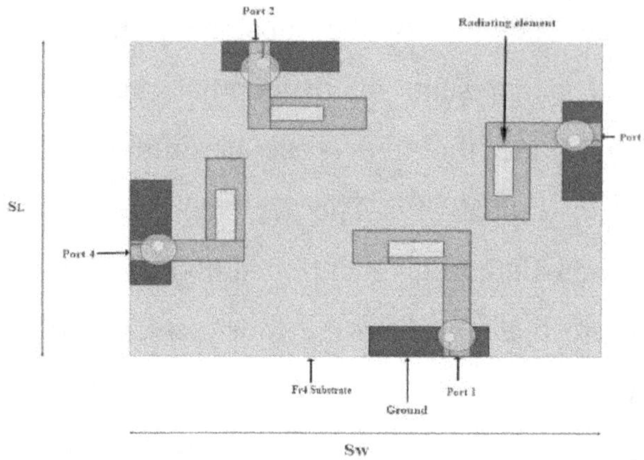

Figure 3.1. The proposed design of the MIMO Antenna

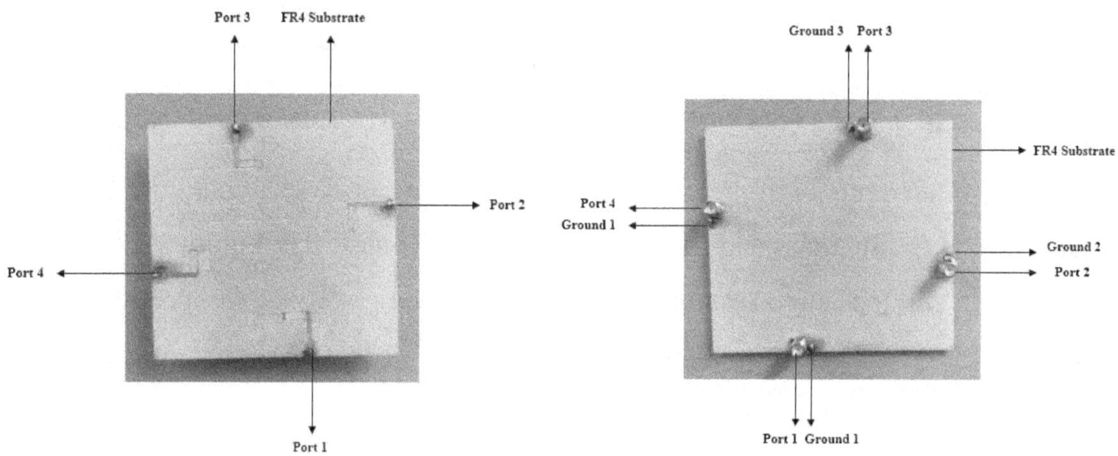

(a) **(b)**

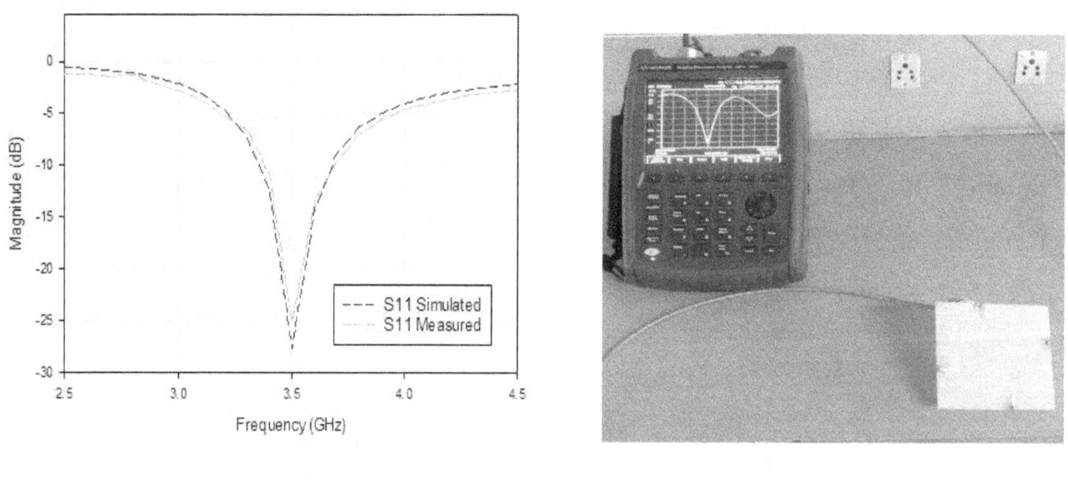

(c) **(d)**

Figure 3.2. Fabricated model (a) Top side, (b) Bottom side, (c) Simulated S- parameter and (d) Measured S-parameter.

Table 3.1. A Comparing Result of Different 5G MIMO Antennas. NG*= Not Given

Reference	Operating Frequency (GHz)	Number of Ports	Measurements (mm^3)	Isolation (dB)	ECC	Peak gain (dBi)	Efficiency (%)
[8]	2.4 GHz	2	22x12x6	-10 dB	0.015	NG	NG
[9]	3.1- 3.8 and 4.8-6 GHz	8	17.85x5x0.8	-10 dB	less than 0.06	NG	65%-75% (LB) and 60%-71% (HB)
[10]	3.3-4.2, 3.3-3.8 and 4.4- 5 GHz	4	120x65x1.6	-18.8 dB	Lower than 0.018	4.71 dBi	NG
[11]	2.4 GHz	4	26x26x0.8	-25 dB	0.02	2.4 dBi	77%
[12]	2.4 GHz	2	30x44x1.6	-35 dB	0.02	1.4 dBi	85%
[13]	3.6 GHz	8	136x68x1	-11.7 dB	Less than 0.1	NG	NG
[14]	2.3 and 4.2 GHz	4	44x44x0.8	-24.8 dB	NG	NG	NG
This Work	**3.5 GHz**	**4**	**96.08x96.08x1.34**	**-27 dB**	**Less than 0.1**	**4.7 dBi**	**87%**

3.1.2 Simulated and Measured Results of the Antenna

The mentioned four-port MIMO antenna system was assembled on a FR4 substrate. Figure 3.2 displays the front and rear surfaces of the constructed MIMO antenna prototype. The following statement includes performance measures for input impedance, mutual coupling between antenna elements, efficiency, radiation characteristics, and envelop correlation coefficient. The intended simulated and measured S-parameters are shown in Figure 3.3. The suggested design includes an estimated normal frequency range at 3.5 GHz and isolations between adjacent and diagonal ports of about -27 dB. This method is illustrated using the antenna arrangement shown in Figure 3.2. This antenna was created on an FR-4 substrate that seems to be 1.34 mm thick. Under each radiating element, four rectangular copper elements were created to represent the antenna's partial ground plane. As seen in the given image, a total of four SMA connectors are used to provide the four port radiators. The identified S-parameters of the suggested configuration in Figure 3.3 are satisfactorily correlated with the simulated results, with just a small variation in the resonant frequency. This can be explained by a variety of factors, including manufacturing method and fault, as well as port/cable connection losses. The isolation between the components of the suggested antenna was examined because the efficiency of antenna diversity is intrinsically tied to the minimal interaction between specific radiating parts. The results of the simulated and experimental isolation among the antenna are shown in Figure 3.4. The isolation values vary from one to the next in the impedance range of the planned antenna. The findings imply that MIMO applications at 3.5 GHz may operate well with isolation greater than -20 dB.

Table 3.2. Optimized Design Parameters of the designed antenna

Parameter	Description	Value
W1	Width of the feedline	1 mm
W2	Width of the patch	2 mm
L1	Length of the patch	18.6 mm
L2	Length of the slot	11mm
L3	Length of the feedline	15 mm
G1	Length of the ground	8.39 mm
G2	Width of the ground	3.6 mm
LS	Length of the substrate	96.08 mm
WS	Width of the substrate	96.08 mm

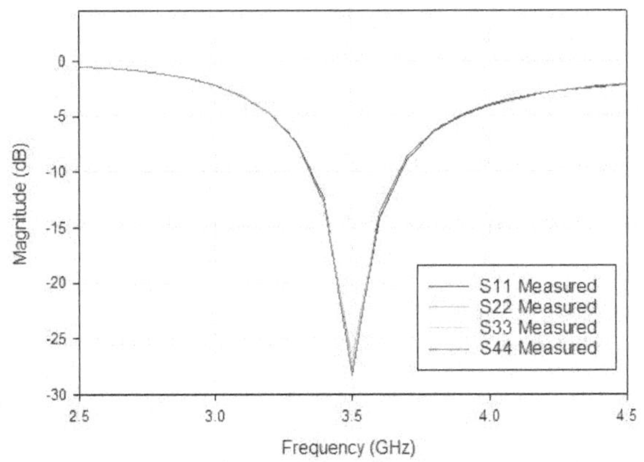

Figure 3.3. Simulated and Measured Return loss

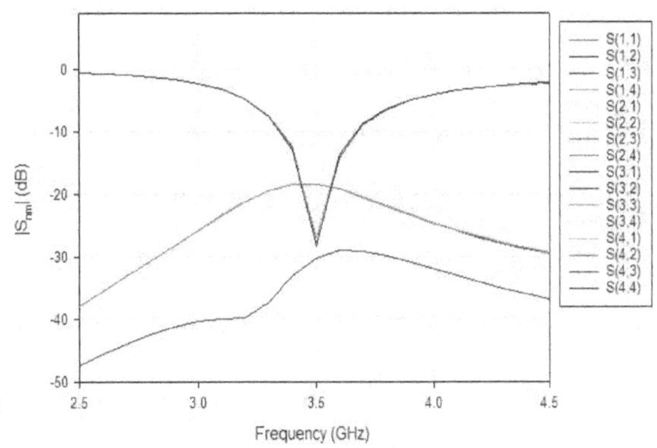

Figure 3.4. Isolation Performance

3.1.3 Electric Current Distribution

The figures above are used to validate the results of the mutualistic link between the radiating components of the antenna design. This may be connected to the antenna's surface current distribution, as displayed in Figure 3.5. To evaluate the impact of connectivity between antenna segments, for instance, one of the four ports should be engaged while the other three ports should be stopped with a 50 Ω load, as illustrated in Figure 3.5. The distance between the ports is crucial for determining how frequently the antennas are separated. There is a significant degree of isolation when the antenna elements are separated by a long distance.

Only port 1 is active, as shown in Figure 3.5(a). On the feeding part, the current concentrations have become more intense as illustrated in Figure 3.5(b), as there is no current associated to the adjoining ports. As a reason, the S12 (or S34) isolation parameter can be used to assess its influence. Only this port has current when port 2 is activated and the other three ports are disconnected with a 50 Ω load. When only port 3 and port 4 are stimulated, a similar condition is portrayed in Figure 3.5(c) and 5(d). As seen in Figures, the surface current is generally intensive over the stimulated port. As a result, the observed and simulated isolations of the suggested antenna in Figures are significant at 10 dB. The E- and the H-Plane at 3.5 GHz, are employed to explore the antenna emission patterns when one port is activated and the others are fitted to a 50Ω load, as illustrated in Figure 3.6.

At the 3.5 GHz resonant frequencies offered by the Ansys HFSS software, both planes of the current MIMO antenna's generated field patterns are shown, as stated. In the case of a standard horn antenna acting as both a transmitter and a power meter, the computed radiation is gathered and completed inside an anechoic chamber to measure the received field. Throughout the measurement process, one port is designated to serve as a receiver, while the remaining ports are saturated with a 50Ω load to prevent signal pickup. For each of the available MIMO design ports, this phase is repeated. The radiation pattern, gain, and reflection coefficients are used to illustrate how well an antenna performs. Two conventional horn antennas and the proposed architecture are tested in an anechoic laboratory to determine the suggested antenna gain, which is a far-field property. The measured and predicted gain at 3.5 GHz are shown in Figure 3.7.

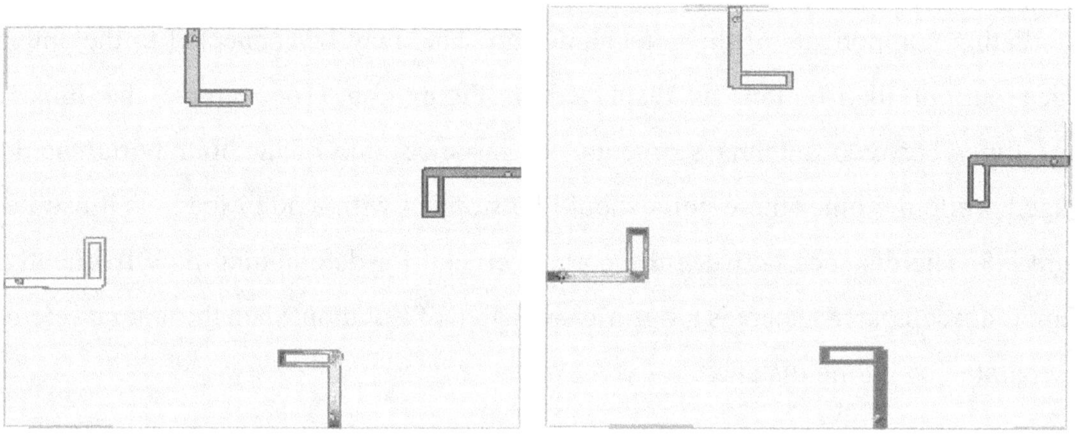

(a) **(b)**

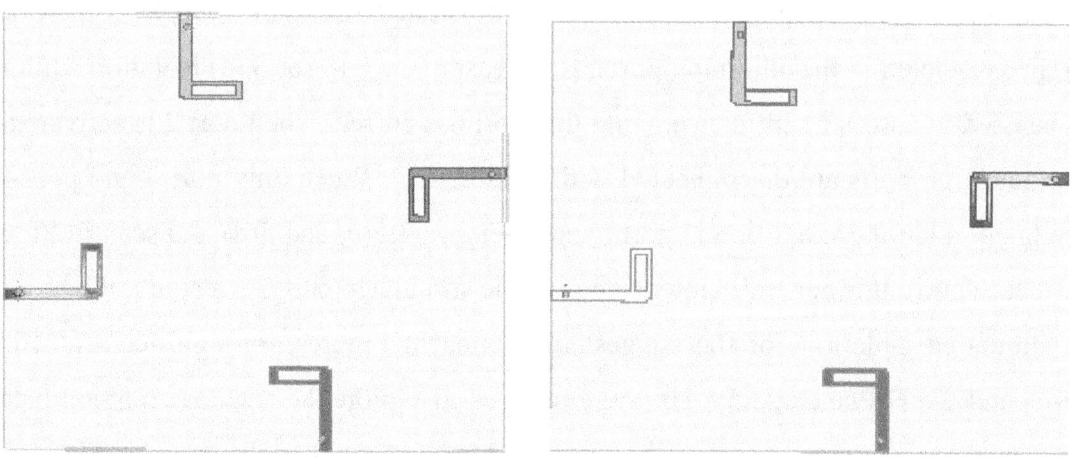

(c) **(d)**

Figure 3.5. Current distribution at 3.5 GHz, Excitation of (a) port 1, (b) port 2, (c) port 3 and (d) port 4

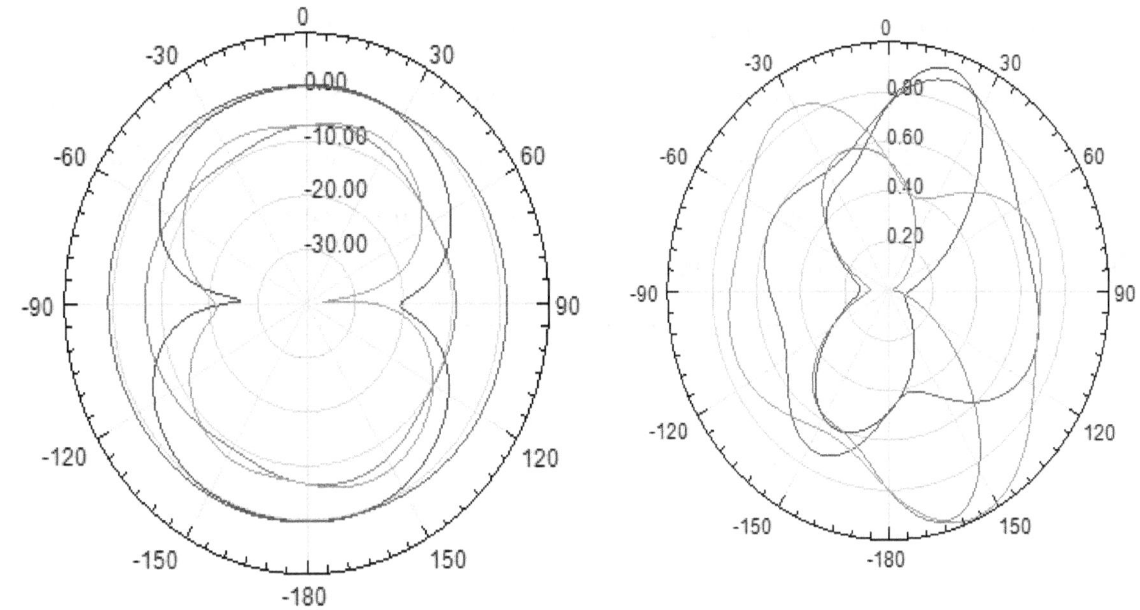

(a) **(b)**

Figure 3.6. (a) Radiation pattern of the projected antenna (both E-plane and H-plane), (b) Gain LHCP and Gain RHCP

Equations (3.3) and (3.4) shows reflection (S11/S22) and transmission coefficients (S12/S21) using source impedance (Z_S) and load impedance (Z_L).

$$\text{Reflection Coefficient} = \frac{Z_L - Z_S}{Z_L + Z_S} \tag{3.3}$$

$$\text{Transmission Coefficient} = \frac{2Z_L}{Z_L + Z_S} \tag{3.4}$$

The radiation pattern can be visualized as a graphical representation of radiated power as a function of a coordinate space, as indicated in equation 3.5. The intensity of the far-field radiation is [4-7]:

$$U = r^2 W_{rad} = B_o F(\theta, \emptyset) \cong \frac{r^2}{\eta} [|E_\theta(r, \theta, \emptyset)|^2 + |E_\theta(r, \theta, \emptyset)|^2] \tag{3.5}$$

where,

U is the intensity of the radiation which is expressed in W/unit solid angle,

W_{rad} is the radiation density having the unit W/m^2,

$E(r, \theta, \phi)$ is the antenna's electric field intensity in the far zone,

E_θ and E_ϕ are the far zone electric field components of antenna,

η is the intrinsic impedance of the medium.

The radiation pattern of the proposed MIMO structure for the frequency 3.5 GHz is shown in Figure 3.6. The image depicts a highly bidirectional and omnidirectional pattern for ϕ- polarized elements (0° and 90°) in the two major planes E and H. The antenna gain is represented in Figure 3.7 & Figure 3.8.

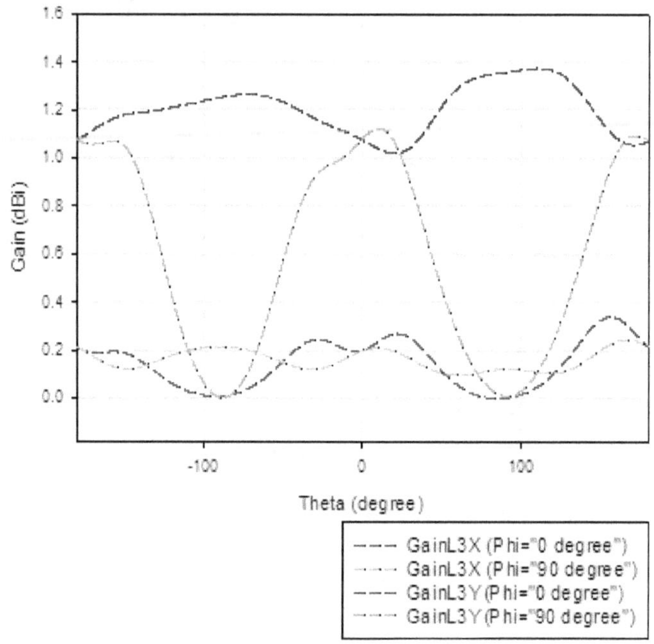

Figure 3.7. Antenna gain at 3.5 GHz

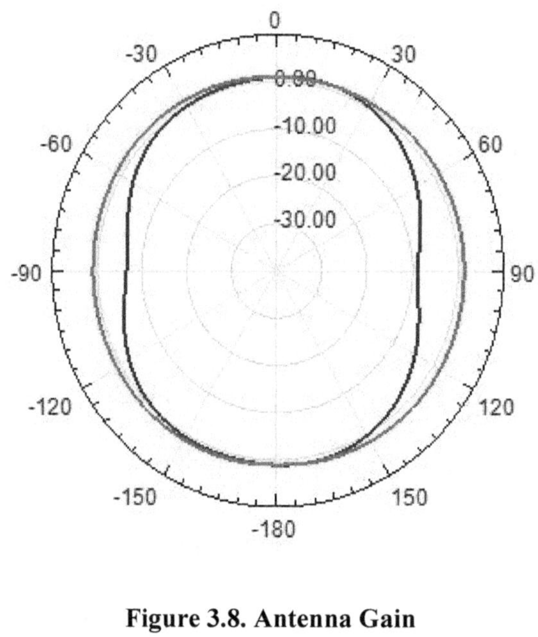

Figure 3.8. Antenna Gain

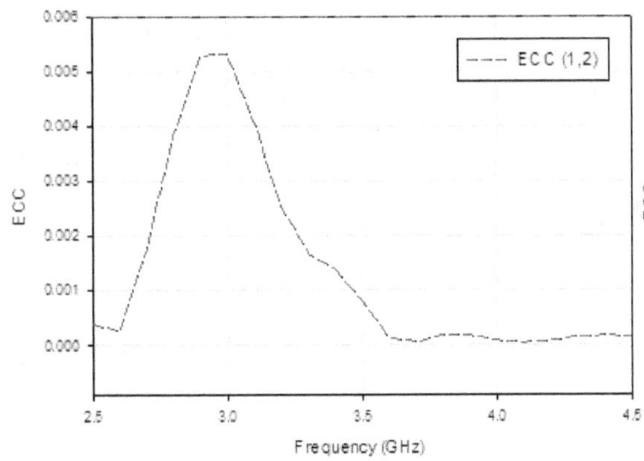

(a)

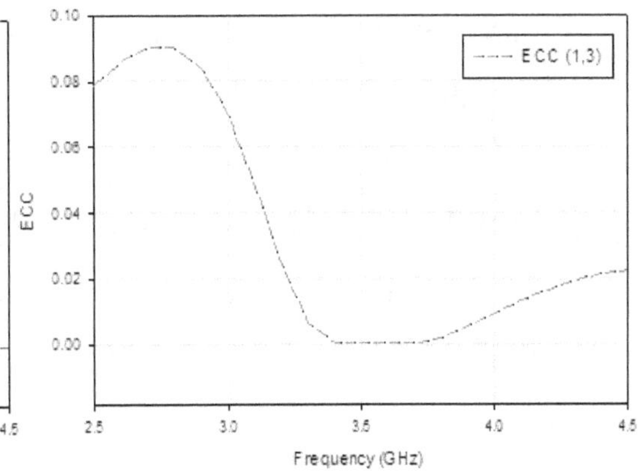

(b)

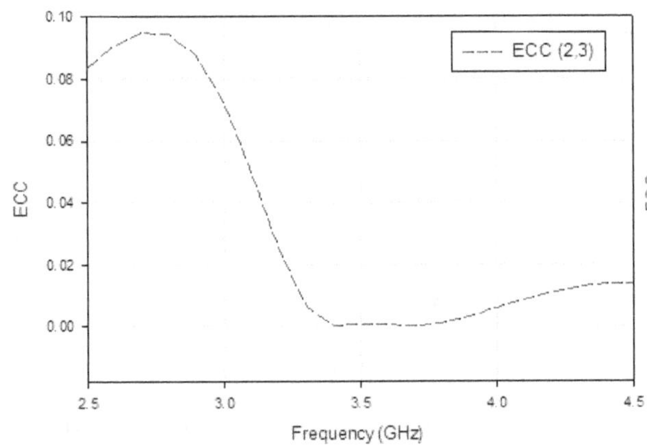

(c)

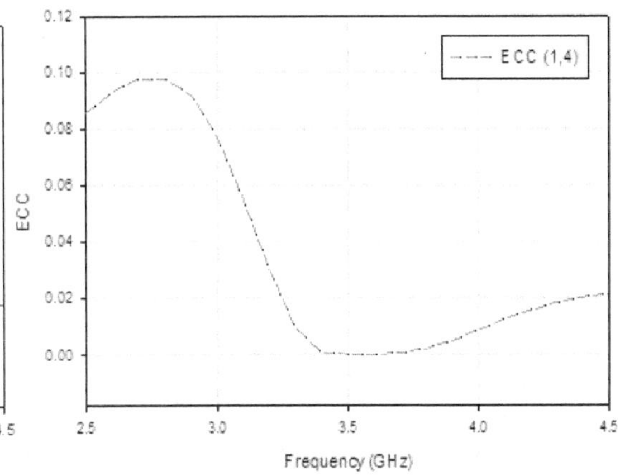

(d)

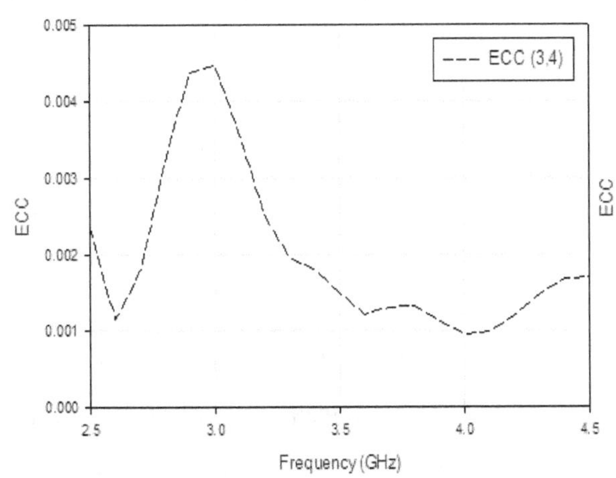

(e)

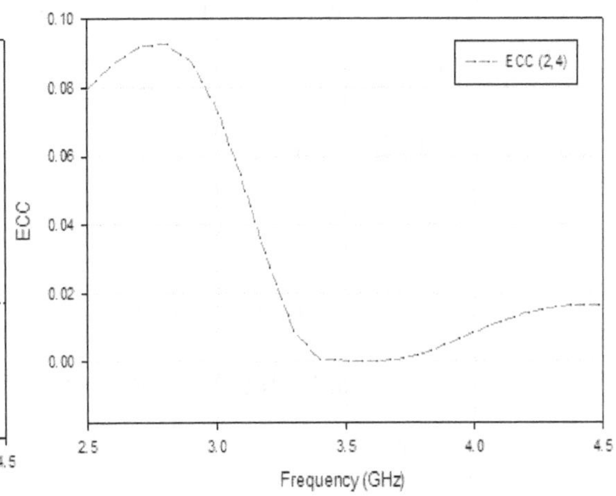

(f)

Figure 3.9. The ECC values of the antenna (a) ECC (1,2), (b) ECC (1,3), (c) ECC (1,4), (d) ECC (2,3), (e) ECC (2,4) and (f) ECC (3,4).

Figure 3.9 displays the calculated ECC seen between ports of the intended work under all conditions. This illustrates how independently the various antenna system components are from one another in terms of radiation pattern. Despite the fact that each component of this antenna system should always be independent of the others, getting a zero value is difficult. The following formula, as demonstrated in Calculations (3.6)-(3.8), can be modified to yield the three ECC measurements in this example by substituting the relevant terms.

$$\rho_{e12}=\frac{|S_{11}S_{12}+S_{12}S_{22}+S_{13}\,S_{32}+S_{14}S_{42}|^2}{(1-(|S_{11}|^2+|S_{12}|^2)+(|S_{13}|^2+|S_{14}|^2))^2} \tag{3.6}$$

$$\rho_{e13}=\frac{|S_{11}S_{13}+S_{12}S_{23}+S_{13}\,S_{33}+S_{14}S_{43}|^2}{(1-(|S_{11}|^2+|S_{12}|^2)+(|S_{13}|^2+|S_{14}|^2))^2} \tag{3.7}$$

$$\rho_{e14}=\frac{|S_{11}S_{14}+S_{12}S_{24}+S_{13}\,S_{34}+S_{14}S_{44}|^2}{(1-(|S_{11}|^2+|S_{12}|^2)+(|S_{13}|^2+|S_{14}|^2))^2} \tag{3.8}$$

Due to the higher potential DG and correlation coefficient, the diversity gain is also regarded as a major diversity characteristic and may be estimated using Equations (3.9)-(3.11). Diversity gain is the term used to describe any diversity technique that enhances the signal-to-interference ratio. The antenna has a greater distribution gain value because it provides better isolation, and vice versa. The equations demonstrate that the diversity gain values for both circumstances are about 10 dB at the functional band and express representations of the diversity gain of the existing antenna.

$$DG_{12} = 10\sqrt{1-|\rho e12|^2} \tag{3.9}$$

$$DG_{13} = 10\sqrt{1-|\rho e13|^2} \tag{3.10}$$

$$DG_{14} = 10\sqrt{1-|\rho e14|^2} \tag{3.11}$$

One of most data that may be transmitted across a communication channel without experiencing a loss is called the channel capacity loss. Adding a greater number of antennas is probably a good way to increase the channel capacity of a system architecture. Meanwhile, the existence of Rayleigh-fading channels will result in a reduction in channel capacity. The following calculation, supposing only the receiving antenna layouts are coupled and the worst-case scenario of a high signal-to-noise ratio, can be used to calculate the channel capacity loss in a 2x2 MIMO system as given in equation (3.12).

$$CCL = -log_2\, det(\varphi^R) \tag{3.12}$$

Where φ^R is referred to as the receiving antenna correlation matrix. It is given by the formula as given in Equation (3.13):

$$\varphi^R = \begin{bmatrix} \rho_{11} & \rho_{12} & \rho_{13} & \rho_{14} \\ \rho_{21} & \rho_{22} & \rho_{23} & \rho_{24} \\ \rho_{31} & \rho_{32} & \rho_{33} & \rho_{34} \\ \rho_{41} & \rho_{42} & \rho_{43} & \rho_{44} \end{bmatrix} \tag{3.13}$$

Where ρ_{ii} and ρ_{ij} is defined as follows given in equations (3.14 & 3.15):

$$\rho_{ii} = 1 - |\textstyle\sum_{n=1}^{N} S_{in} S_{ni}| \text{ for i, j=1,2,3 or 4} \tag{3.14}$$

$$\rho_{ij} = 1 - |\textstyle\sum_{n=1}^{N} S_{in} S_{nj}| \text{ for i, j=1,2,3 or 4} \tag{3.15}$$

The power differential between a MIMO antenna under test and the reference standard MIMO antenna needed to achieve a given performance is known as multiplexing efficiency (ME). The TARC factors in coupling and random signal configurations to calculate the MIMO antenna's effective operational bandwidth in addition to the diversity characteristics mentioned above. The TARC and ME of the suggested design, respectively, are much below -10 dB and much less than -1 dB across the impedance bandwidth. These graphs demonstrate how well-suited the antenna configuration is for MIMO applications.

The value of TARC and ME are considered using S-parameters, as presented in the below equations (3.16) and (3.17).

$$\text{TARC} = \sqrt{\frac{(S_{11}+S_{12})^2+(S_{21}+S_{22})^2}{2}} \tag{3.16}$$

$$\text{ME} = \sqrt{\eta_i \eta_j \left(1 - |\rho_{ij}|^2\right)} \tag{3.17}$$

Where η_i and η_j are the total efficiency of ith and jth element, ρ_{ij} is correlation coefficient between i^{th} and j^{th} element.

3.2 CONCLUSION

A smaller four-element MIMO antenna for the 5G application is suggested and shown. The P-shaped radiating elements are produced on a basic FR-4 material. The proposed antenna's envelope dimension is minimal. Both calculations and fabrication have been done for the mentioned antenna design. Overall estimates and testing indicated a return loss of -27 dB at the specified 3.5 GHz range. The envelope correlation coefficient, antenna gain, performance, and channel capacity loss are some further findings. The proposed antenna was also put to the test, and it was discovered to be more efficient than the free space antenna design. This has led to evidence that the recommended antenna works effectively for applications in both mobile and satellite communications as well as in healthcare.

References

[1]. Jensen M.A and Wallace J.W," A Review of antennas and propagation for MIMO wireless communications", *IEEE Trans. Antennas Propag.*,52, 2810–2824, **2004**.

[2]. Liao W.J, Hsieh C.Y, Dai B.Y and Hsiao B.R, "Inverted-F/slot integrated dual-band four-antenna system for WLAN access points", *IEEE Antenn. Wirel. Pr.*, 14, 847–850, **2014**.

[3]. Iqbal A, Saraereh O.A, Ahmad A.W and Bashir S, "Mutual coupling reduction using F-shaped stubs in UWB-MIMO antenna", *IEEE Access*, 6, 2755–2759, **2017**.

[4]. Abdullah M, Kiani S.H, Iqbal A, "Eight Element Multiple-Input Multiple-Output (MIMO) Antenna for 5G Mobile Applications", *IEEE Access*, 7, 134488–134495, **2019**.

[5]. Abdullah M, Kiani S.H, Abdulrazak L.F, Iqbal A, Bashir M, Khan S and Kim S, "High-Performance Multiple-Input Multiple-Output Antenna System for 5G Mobile Terminals", *Electronics*, 8, 1090, **2019**.

[6]. Iqbal A, Basir A, Smida A, Mallat N.K, Elfergani I, Rodriguez J and Kim S," Electromagnetic bandgap backed millimeter-wave MIMO antenna for wearable applications", *IEEE Access*, 7, 111135–111144, **2019**.

[7]. Ahmad M.S, Mohyuddin W, Choi H.C and Kim K.W," 4x4 MIMO antenna design with folded ground plane for 2.4 GHz WLAN applications", *Microw. Opt. Technol. Lett.*, 60, 395–399, **2018**.

[8]. Alibakhshikenari M, Khalily M, Virdee B.S, See C.H, Abd-Alhameed R.A and Limiti, E," Mutual-Coupling Isolation Using Embedded Metamaterial EM Bandgap Decoupling Slab for Densely Packed Array Antennas", *IEEE Access* ,7, 51827–51840, **2019**.

[9]. Yang L, Li T and Yan S," Highly compact MIMO antenna system for LTE/ISM applications", *Int. J. Antennas Propag*, 714817, **2015**.

[10]. Issa Elfergani, Maryam Sajedin," Low profile and closely spaced four element MIMO antenna for wireless body area networks", *Electronics*, **2020**.

[11]. Kiem N.K, Phuong H.N.B, Hieu Q.N, Chien D.N," A novel metamaterial MIMO antenna with high isolation for WLAN applications", *Int. J. Antennas Propag.*, **2015**.

[12]. Kiem, N. K., H. N. B. Phuong, Q. N. Hieu, and D. N. Chien, "A novel metamaterial MIMO antenna with high isolation for WLAN applications," *Int. J. Antennas Propag.*, Vol. 2015, Article ID 851904, 9 pages, **2015**.

[13]. Singh H. S., G. K. Pandey, P. K. Bharti, and M. K. Meshram, "A compact dual-band diversity antenna for WLAN applications with high isolation," *Microw. Opt. Technol. Let.*, Vol. 57, 906–912, **2015**.

[14]. Likhitha T., S. Ashok Kumar, and T. Shanmuganantham, "Design of compact four port MIMO antenna using SRR ring for high isolation," *Proceedings of the International Conference on Antenna Testing and Measurement Society (ATMS 2018)*, 1–4, Pune, India, February 6–7, **2018**.

Chapter IV

CHAPTER IV

A SLOTTED COMPACT FOUR-PORT TRUNCATED GROUND STRUCTURED MIMO ANTENNA FOR SUB-6 3.4 GHz 5G APPLICATION

A Novel high performance four-port multiple input multiple output antenna is suggested for 5G application functioning at a 3.4 GHz band. The antenna design measures an inclusive volume of 32mm x 32mm x 0.8mm^3. The broad frequency bandwidth, excellent gain, decreased inter-element gap and effective isolation within the MIMO components of the proposed system are clearly novel. Each antenna in the four-element MIMO system has been situated orthogonally to the others while maintaining a small size and a good result. The antenna has exceptional average total efficiency in the 5G Sub-6 GHz spectrum and is in good agreement with the measured results. It also offers a high realized gain when compared to prior MIMO antennas. The antenna is having a high impedance matching whose isolation is about -28 dB, envelope correlation coefficient is computed and is smaller than 0.10, channel capacity loss average value of less than 0.2 bits per second per hertz and the diversity gain of about 10 dB. The typical peak realized antenna gain of the offered MIMO antenna is also delivered with a high radiation efficiency at the frequency of 3.4 GHz. The reflection coefficient, mutual coupling, radiation pattern, current distribution and gain of antennas are all measured and explained. The design is having a compact high volume and adequate bandwidth with good accomplished gain making the antenna a very strong for 5G application.

4.1 CONCEPT OF THE DESIGNED ANTENNA

4.1.1 Single Antenna Design and Analysis

This section goes through the proposed single antenna structure and design strategy in detail. Furthermore, the scattering characteristics, gain, and overall efficiency of the measured and simulated results are evaluated. The antenna model is built on a FR4 Epoxy substrate with a low dielectric constant and 0.8 mm thickness. Ansys HFSS electromagnetic simulation software and the CST Studio package 2019 were used to develop and model this design.

4.1.2 Single Antenna Geometry and its Design Procedures

The technique shows the proposed layout and design concept for a single-element antenna. According to the well-established mathematical equation, given in (4.1 & 4.2) the antenna consists of a radiating patch with a feedline and a copper ground plane, and it resonates with a tolerable bandwidth.

$$P_W = \frac{C}{2f_r}\sqrt{\frac{2}{\gamma_r+1}} \tag{4.1}$$

$$P_L = \frac{C}{2f_r\sqrt{\gamma_{eff}}} - 2\Delta L \tag{4.2}$$

Where, the patch length and the width are represented as P_L and P_w, C denotes the speed of light, γ_r is the dielectric constant of the given substrate, γ_{eff} indicates the effective dielectric value of the radiating patch and ΔL denotes the change in patch length. The suggested antenna's impedance bandwidth and impedance matching are enhanced by moving the feed line's location to the right [1-4]. The antenna covered the Sub-6 GHz frequency range in 5G during this phase and displayed exceptional operational bandwidth of 3.4 GHz. The radiating patch's corner square plots are etched in the following step. The suggested design's front and bottom viewpoints of both the single and overall antenna prototypes are shown in Figure 4.1, along with the final optimized intended design characteristics.

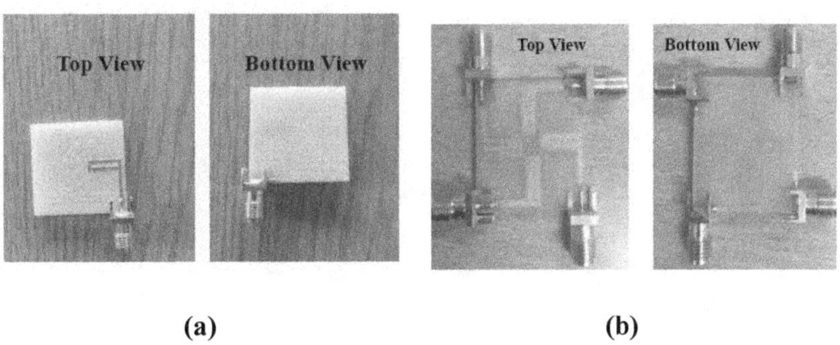

Figure 4.1. Fabricated prototype of (a) the single antenna element and (b) the overall antenna

4.1.3 Design of Overall antenna geometry

Figure 4.2 explains the planned antenna, which comprises of four P-shaped monopole antenna elements. On a FR-4 dielectric substrate, the intended 3.4 GHz MIMO antenna is built. With a total substrate size of 32x32x0.8 mm^3, the suggested configuration of the four division MIMO antenna is recognized as a lowest possible design in 5G antennas. As depicted in Figure 4.2, the microstrip design is considered as a particular 50Ω microstrip line that is constructed and created on a FR-4 material. At first, only one of the antenna elements were considered. At the required frequency, the antenna's dimensions were optimized as 3.4 GHz. Figure 4.4 [a] shows a simulated reflection coefficient of the single antenna element. Figure 4.4 [b] shows the antenna's simulated and measured return loss. As seen in the figures, the simulated and measured findings had nearly the same resonance frequency as shown in Figure 4.4 [c] and [d]. The constructed prototypes bandwidth is lower than the derived prototypes, but it is still greater than the needed band. With a separation spacing, the four p-shaped portions are symmetrically positioned. The absolutely tiny ground area is engraved on the lower side of the material when this nomenclature is used. Each radiating element's ground plane was split into four rectangular sections made below it.

The conceptual outlooks of the assessed radiating element which is in the front part and partial ground which is on the back side, as well as the optimum size, are shown in Figure 4.2. With the help of Ansys HFSS, the current design is simulated. Table 4.1 shows the comparison of different 5G MIMO antennas and Table 4.2 lists the geometrical parameters in detail, along with their sizes.

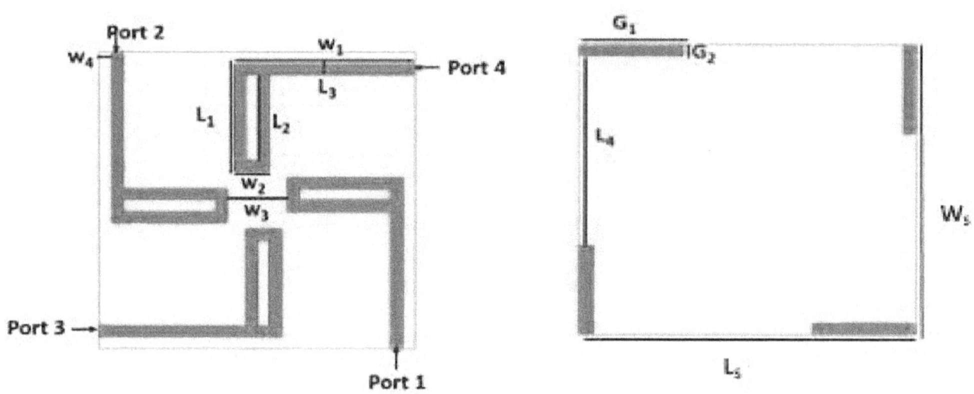

Figure 4.2. Designed antenna with Partial ground

Table 4.1. A Comparing Result of Different 5G MIMO Antennas NG*= Not Given

Reference	Operating Frequency (GHz)	Number of Ports	Measurements (mm^3)	Isolation (dB)	ECC	Peak gain (dBi)	Efficiency (%)
[8]	2.4 GHz	2	22x12x6 mm^3	-10 dB	0.015	NG	NG
[9]	3.1- 3.8 and 4.8-6 GHz	8	17.85x5x0.8 mm^3	-10 dB	less than 0.06	NG	65%-75% (LB) and 60%-71% (HB)
[10]	3.3-4.2, 3.3-3.8 and 4.4-5 GHz	4	120x65x1.6 mm^3	-18.8 dB	Lower than 0.018	4.71 dBi	NG
[11]	2.4 GHz	4	26x26x0.8 mm^3	-25 dB	0.02	2.4 dBi	77%
[12]	2.4 GHz	2	30x44x1.6 mm^3	-35 dB	0.02	1.4 dBi	85%
[13]	3.6 GHz	8	136x68x1 mm^3	-11.7 dB	Less than 0.1	NG	NG
[14]	2.3 and 4.2 GHz	4	44x44x0.8 mm^3	-24.8 dB	NG	NG	NG
This work	**3.4 GHz**	**4**	**32 x 32 x 0.8** mm^3	**-28 dB**	**Less than 0.1**	**3.2 dBi**	**82%**

Table 4.2. Optimized Design Parameters of the designed antenna

Parameter	Description	Value
W1	Width of the feedline	1mm
W2	Width of the patch	2mm
L1	Length of the patch	13mm
L2	Length of the slot	11mm
L3	Length of the feedline	15mm
G1	Length of the ground	8mm
G2	Width of the ground	1mm
LS	Length of the substrate	32mm
WS	Width of the substrate	32 mm

4.1.4 Equivalent Circuit Model

An equivalent lumped circuit model is constructed utilising an advanced design methodology in order to examine the true characteristics of the proposed four elements MIMO antenna. Figure 3 depicts the similar circuit model. The parallel RLC resonant circuit represents each monopole radiator of the MIMO antenna system. A 50 Ω terminal excites each monopole. An LC circuit represents the related coupling between the MIMO elements. To obtain the optimum circuit values, the analogous circuit is fine-tuned. Table 4.3 shows the optimised circuit parameters. Figures compares the s-parameters of the circuit and the EM model. This can be seen that the presented curves have good agreement.

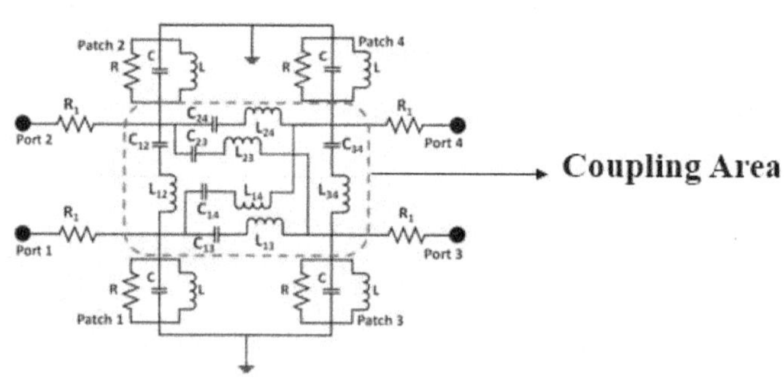

Figure 4.3. Equivalent circuit model of the MIMO antenna

Table 4.3. Optimized Circuit Parameters

Component	Value	Component	Value
R_1	1 Ω	R	9.4Ω
L	10.50 nH	C	6.1 pF
L_{12}	1.48 nH	C_{12}	2.22 pF
L_{13}	7.60 nH	C_{13}	6.38 pF
L_{14}	2.20 nH	C_{14}	1.48 pF
L_{23}	1.5 nH	C_{23}	1.49 pF
L_{24}	11.28 nH	C_{24}	5.65 pF
L_{34}	16.92 nH	C_{34}	5.4 pF

4.2 RESULTS AND DISCUSSION

4.2.1 Simulated and Measured Results

On a FR4 substrate, the desired four-port MIMO antenna system was set up. The foremost and the backmost surfaces of the manufactured MIMO antenna prototype are presented. Input impedance, mutual coupling between antenna elements, efficiency, radiation characteristics, and envelop correlation coefficient are all performance measures included in the following statement. Figure 4.4 depicts the designed simulated and measured S-parameters. At 3.4 GHz, the projected design has a modelled normal frequency range, with isolations of around -28 dB between adjoining and diagonal ports. This approach is demonstrated using the antenna configuration. This antenna was produced on a 0.8 mm thick FR-4 substrate. Four rectangular copper elements were fabricated beneath each radiating element to simulate the antenna's ground area. Totally there are four connectors used to feed the four port radiators, as indicated in the given figure. With just a slight shift in the resonant frequency, the observed S-parameters of the proposed configuration are in satisfactory correlation with the simulated findings, which can be attributed by a range of factors including manufacturing method and error, along with port/cable connection losses. Because the least interaction between specific radiating elements is necessarily tied to the antenna diversity efficiency, the isolation seen between elements of the recommended antenna was researched and evaluated.

In the figure 4.4, it shows the simulated and experimental isolation results amongst the antenna. In the projected antenna's impedance band, the isolation values fluctuate from one to the following. The results suggest that MIMO applications around 3.4 GHz can work well, with isolation stronger than -20 dB.

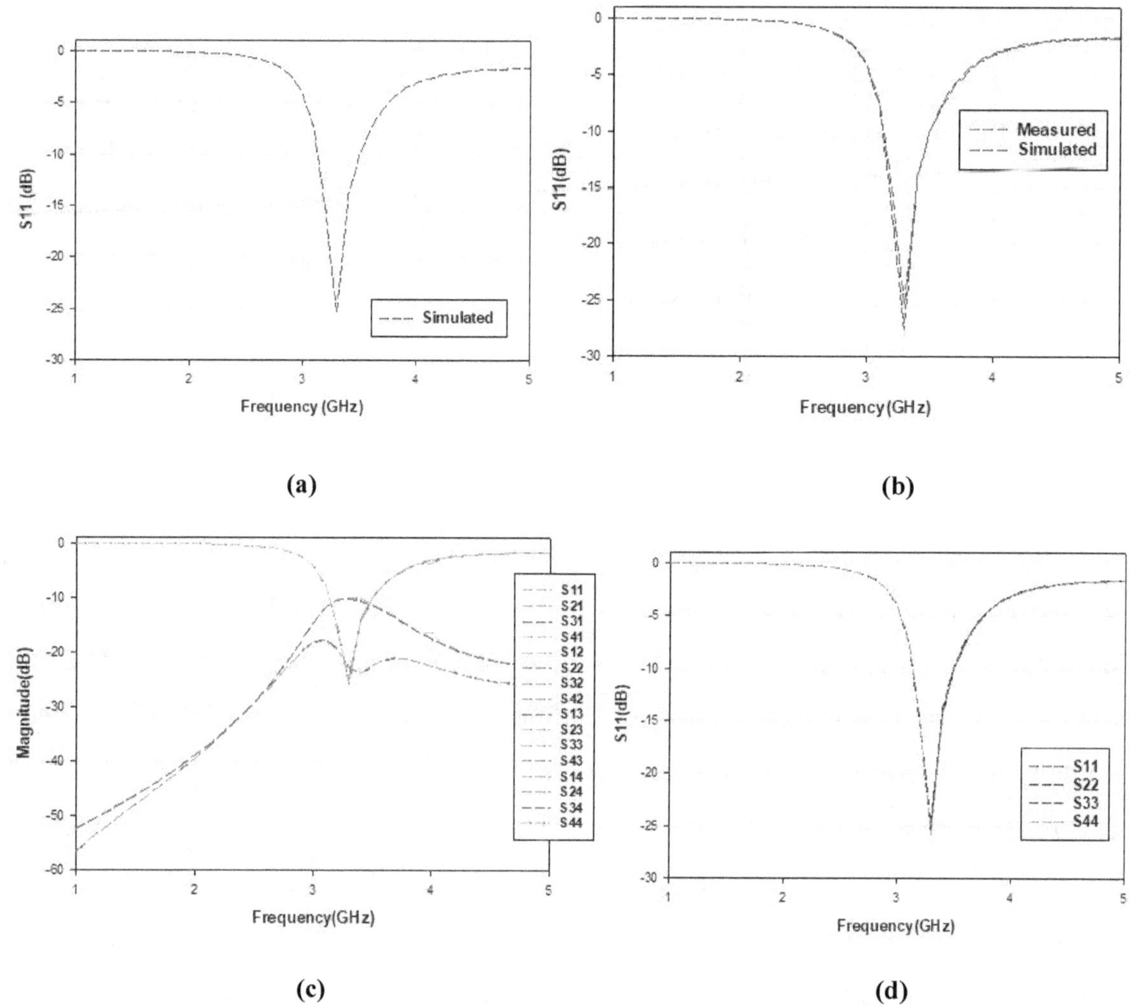

Figure 4.4. (a) Simulated Reflection coefficient of a single antenna, (b) Simulated and Measured S-parameter, (c) Simulated results of a proposed MIMO antenna with all ports, (d) Isolation Performance of the suggested antenna

4.2.2 Electric Current Distribution

The outcomes of the mutual connection between the radiating features of the antenna design are validated using the figures above. This can be related to the surface current distribution of the antenna as shown in Figure 4.5. For instance, one of the four ports should be activated to investigate the effect of connectivity between antenna segments using this technique, while the other three ports should indeed be stopped with a 50 Ω load, as shown in Figure 4.5. When calculating how often the antennas are apart, the distance between the ports is essential. When the antenna elements are divided by a large distance, there is a substantial level of isolation. Only port 1 is active, as shown in Figure 4.5 [a]. On the feeding part, the current concentrations have become more intense as illustrated in Figure 4.5 [a], as there is no current associated to the adjoining ports. As a reason, the S12 (or S34) isolation parameter can be used to assess its influence. Only this port has current when second port is activated and the other ports are disconnected with a 50 Ω load. When port 3 and port 4 are stimulated, a similar condition is portrayed in Figure 4.5[c], 3[d]. As seen in Figure, the surface current is generally intensive over the stimulated port. As a result, the observed and simulated isolations of the suggested antenna in Figures are significant at 10 dB.

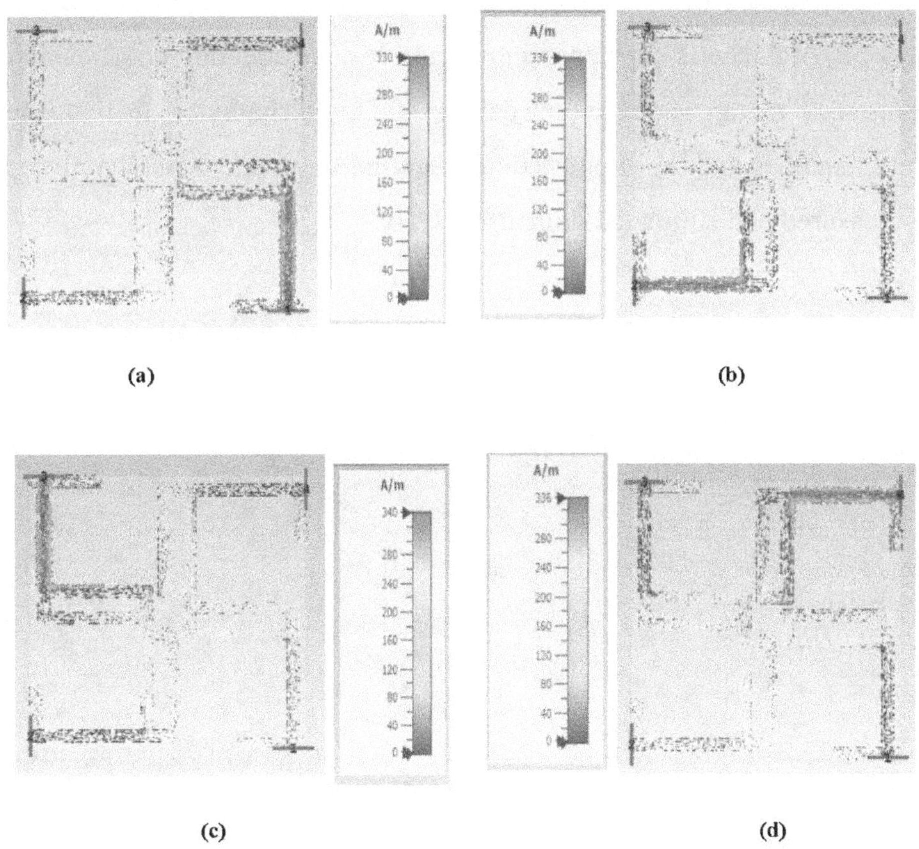

(a) (b)

(c) (d)

Figure 4.5. Surface Current distribution at 3.4 GHz, Excitation of (a) port 1, (b) port 2, (c) port 3 and (d) port 4

4.2.3 Radiation Pattern and Gain

The E- and the H-Plane at 3.4 GHz, are employed to explore the antenna emission patterns when one port is activated and the others are fitted to a 50-ohm load, as illustrated in Figure 4.6. Both the planes are generated field patterns of the current MIMO antenna are displayed at the 3.4 GHz resonant frequencies provided by the Ansys HFSS software, as indicated. The radiation pattern of the proposed MIMO structure for the frequency 3.4 GHz is shown in Figure 4.6. The image depicts a highly bidirectional and omnidirectional pattern for ϕ- polarized elements (0° and 90°) in the two major planes E and H. The computed radiation is acquired and accomplished inside an anechoic chamber to measure the received field when a conventional horn antenna is used as a transmitter and a power meter. One port is specified to perform as a receiver throughout the measurement technique, while the subsequent ports are saturated with a 50 Ω load to inhibit signal pickup. This step is repeated for each of the MIMO design ports that have been offered. The antenna performance is expressed in relation to the reflection coefficients, gain, and radiation pattern. The suggested antenna gain is a far-field characteristic that is tested using two standard antennas and the suggested architecture in an anechoic chamber. Figure 4.7 shows the measured and simulated gain at 3.4 GHz.

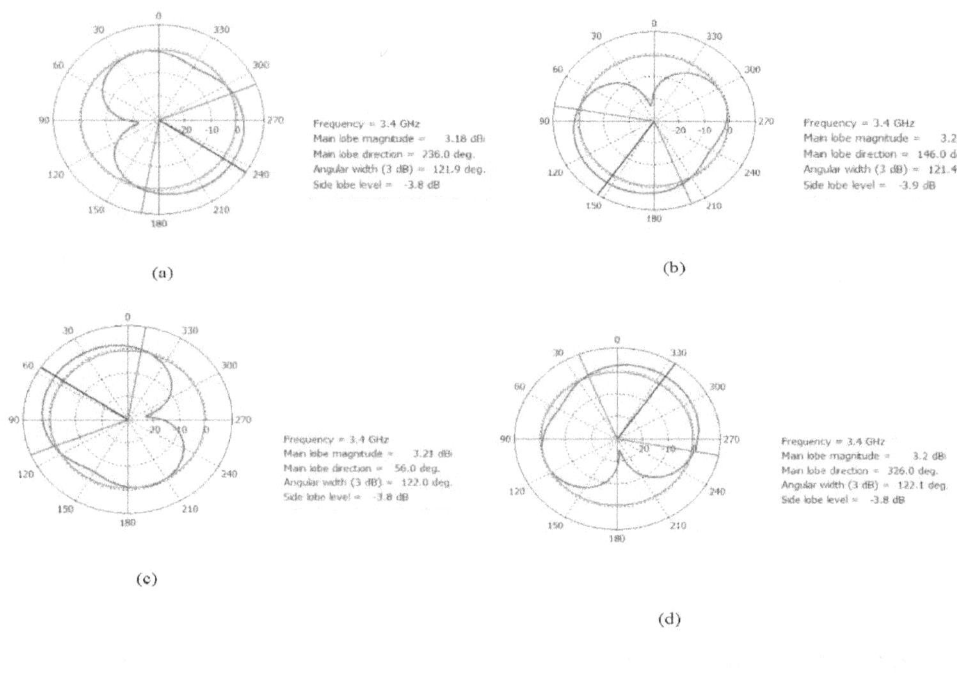

Figure 4.6. Radiation pattern of the projected antenna

Equations (4.3) and (4.4) shows reflection (S11/S22) and transmission coefficients (S12/S21) using source impedance (ZS) and load impedance (ZL).

$$\text{Reflection Coefficient} = \frac{Z_L - Z_S}{Z_L + Z_S} \quad (4.3)$$

$$\text{Transmission Coefficient} = \frac{2Z_L}{Z_L + Z_S} \quad (4.4)$$

As shown in equation 4.5, the radiation pattern can be shown as a graphical representation of generated power as a function of a coordinate space. The far-field radiation intensity is [5-7]:

$$U = r^2 W_{rad} = B_o F(\theta, \emptyset) \cong \frac{r^2}{\eta}[|E_\theta(r,\theta,\emptyset)|^2 + |E_\theta(r,\theta,\emptyset)|^2] \quad (4.5)$$

Where,

U is the intensity of the radiation which is expressed in W/unit solid angle,

$Wrad$ is the radiation density having the unit W/m^2,

$E\ (r, \theta, \phi)$ is the antenna's electric field intensity in the far zone,

E_θ and E_ϕ are the far zone electric field components of antenna,

η is the intrinsic impedance of the medium.

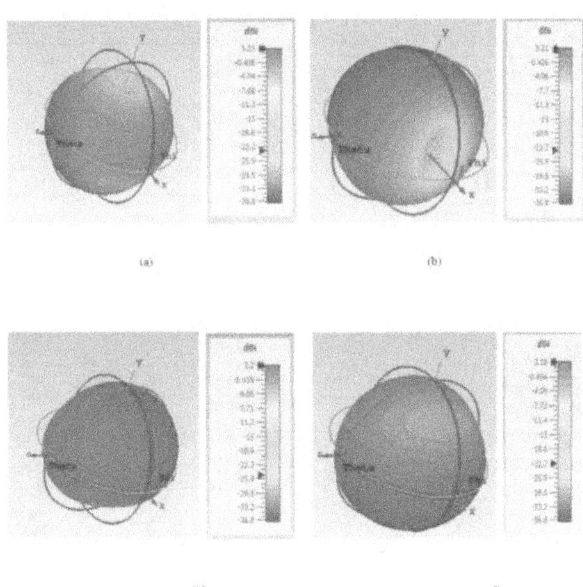

Figure 4.7. 3D Polar plot of the proposed antenna

4.2.4 MIMO Antenna Performances

The determined ECC seen between ports of the designed work in all the circumstances, is shown in Figure 4.8. In perspective of radiation pattern, this depicts how independent the various components of the antenna system are. Each element in this antenna system should always be separate of the others; yet, in reality, achieving a zero value is challenging. The three ECC readings in this case may be obtained by replacing the necessary terms into the following formula, as shown in Calculations (4.6) - (4.8):

$$\rho_{e12}=\frac{|S_{11}S_{12}+S_{12}S_{22}+S_{13}\,S_{32}+S_{14}S_{42}|^2}{\left(1-(|S_{11}|^2+|S_{12}|^2\,)+(|S_{13}|^2+\,|S_{14}|^2)\right)^2} \tag{4.6}$$

$$\rho_{e13}=\frac{|S_{11}S_{13}+S_{12}S_{23}+S_{13}\,S_{33}+S_{14}S_{43}|^2}{\left(1-(|S_{11}|^2+|S_{12}|^2\,)+(|S_{13}|^2+\,|S_{14}|^2)\right)^2} \tag{4.7}$$

$$\rho_{e14}=\frac{|S_{11}S_{14}+S_{12}S_{24}+S_{13}\,S_{34}+S_{14}S_{44}|^2}{\left(1-(|S_{11}|^2+|S_{12}|^2\,)+(|S_{13}|^2+\,|S_{14}|^2)\right)^2} \tag{4.8}$$

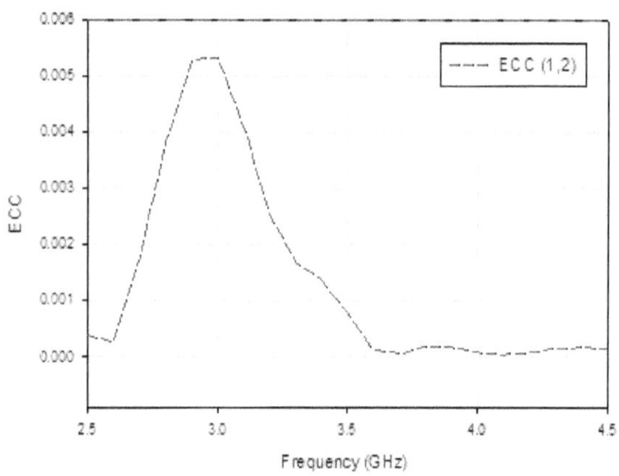

(a)

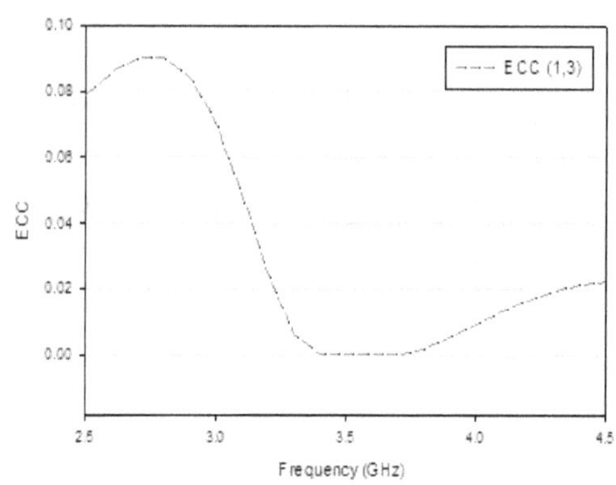

(b)

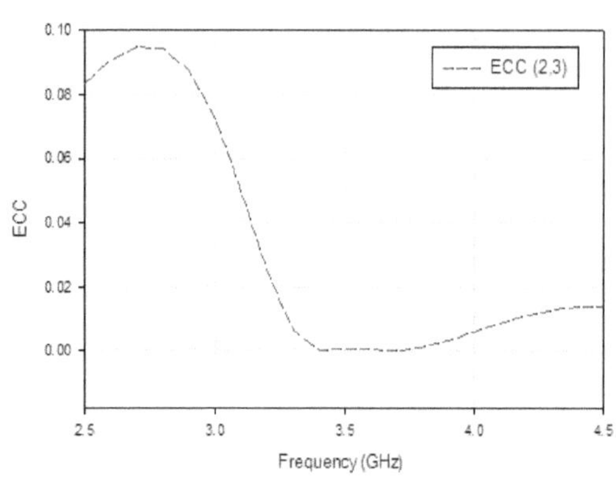

(c)

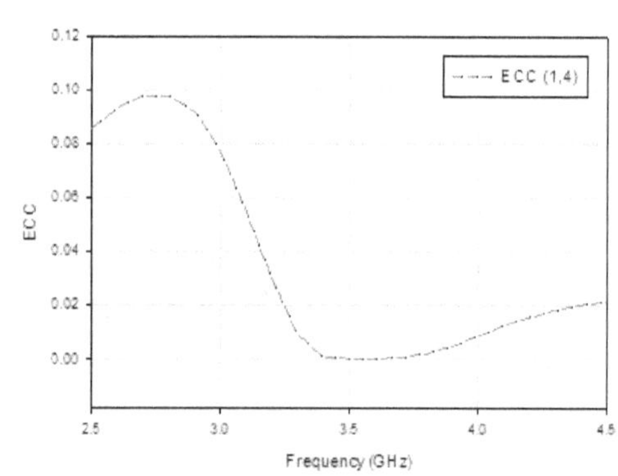

(d)

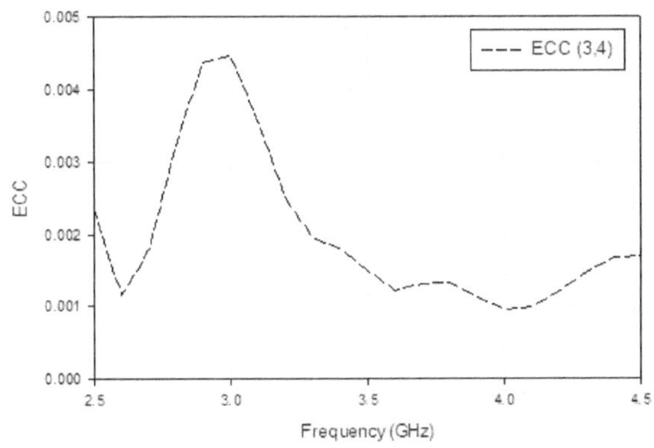

(e)

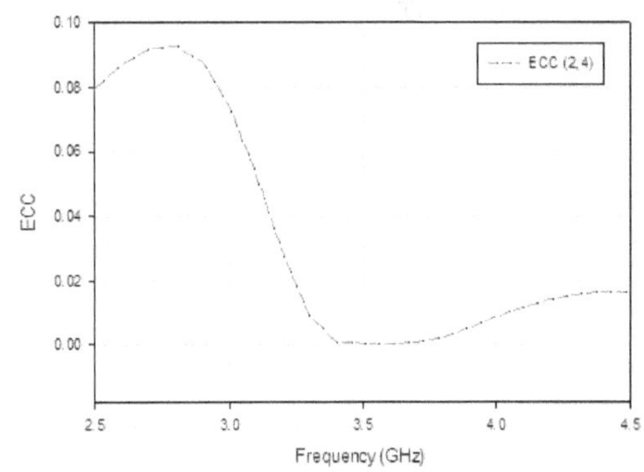

(f)

Figure 4.8. The ECC values of the antenna (a) ECC (1,2), (b) ECC (1,3), (c) ECC (1,4), (d) ECC (2,3) (e) ECC (2,4) and (f) ECC (3,4)

The diversity gain is also considered as a significant diversity characteristic, which may be calculated using Equation (4.9- 4.11) on account of increased possible DG and correlation coefficient. Any diversity method that optimizes the signal-to-interference ratio is described as diversity gain. The antenna gives excellent isolation because it has a higher distribution gain value, and vice versa. The Equations expresses the representations of the diversity gain of the current antenna and also shows that the DG values for both scenarios are around 10 dB at the functional band.

$$DG_{12} = 10\sqrt{1 - |\rho e12|^2} \quad (4.9)$$

$$DG_{13} = 10\sqrt{1 - |\rho e13|^2} \quad (4.10)$$

$$DG_{14} = 10\sqrt{1 - |\rho e14|^2} \quad (4.11)$$

The channel capacity loss given in equation (4.12) is the maximum amount of data that may be carried without loss through a communication channel. A system design channel capacity can presumably be improved by increasing the number of antennas. The occurrence of Rayleigh-fading channels, meanwhile, will lead to a decrease in channel capacity. The channel capacity loss in a 2x2 MIMO system can be obtained from the following equation, assuming only the receiving antenna layouts are correlated and the worst-case scenario of a high signal-to-noise ratio.

$$CCL = -log_2\, det(\varphi^R) \quad (4.12)$$

Where φ^R is referred to as the receiving antenna correlation matrix.

It is given by the formula 4.13 as given below:

$$\varphi^R = \begin{bmatrix} \rho_{11} & \rho_{12} & \rho_{13} & \rho_{14} \\ \rho_{21} & \rho_{22} & \rho_{23} & \rho_{24} \\ \rho_{31} & \rho_{32} & \rho_{33} & \rho_{34} \\ \rho_{41} & \rho_{42} & \rho_{43} & \rho_{44} \end{bmatrix} \tag{4.13}$$

Where ρ_{ii} and ρ_{ij} is defined as follows given in equation 4.14 & 4.15

$$\rho_{ii} = 1 - \left|\sum_{n=1}^{N} S_{in} S_{ni}\right| \text{ for i, j=1,2,3 or 4} \tag{4.14}$$

$$\rho_{ij} = 1 - \left|\sum_{n=1}^{N} S_{in} S_{nj}\right| \text{ for i, j=1,2,3 or 4} \tag{4.15}$$

Multiplexing efficiency is the difference in power required for a MIMO antenna (on test) to achieve a specific throughput compared to the reference standard MIMO antenna (ME). In combination to the diversity parameters described above, the TARC determines the MIMO antenna's efficient operating bandwidth by factoring coupling and random signal configurations. The suggested design's TARC and ME are considerably below -10 dB and less than -1 dB across impedance bandwidth, correspondingly. These figures show that the antenna arrangement presented is well appropriate for MIMO applications.

The value of TARC and ME are considered using S-parameters, as presented in the below equations (4.16 & 4.17).

$$\text{TARC} = \sqrt{\frac{(S_{11}+S_{12})^2+(S_{21}+S_{22})^2}{2}} \tag{4.16}$$

$$\text{ME} = \sqrt{\eta_i \eta_j \left(1 - |\rho_{ij}|^2\right)} \tag{4.17}$$

Where η_i and η_j are the total efficiency of ith and jth element, ρ_{ij} is correlation coefficient between ith and jth element.

4.3 CONCLUSION

For the 5G application, a miniature four-element MIMO antenna is proposed and illustrated. On a minimal FR-4 material, the P radiating elements are generated. The suggested antenna has a small envelope dimension. The referred antenna design has been both calculated and fabricated. At the specified band of 3.4 GHz, both calculations and testing showed with a return loss of -28 dB. Other results include the envelope correlation coefficient, antenna gain, efficiency, and channel capacity loss, among others. The proposed antenna was also tested, and its efficiency was found to be more effective when opposed to the free space antenna design. As a result, the suggested antenna has been demonstrated to be effective to mobile and satellite communications as well as medical applications.

References

[1]. Jensen M. A. and J. W. Wallace, "A Review of antennas and propagation for MIMO wireless communications," *IEEE Trans. Antennas Propag.*, Vol. 52, 2810–2824, **2004**.

[2]. Elfergani I, A. S. Hussaini, J. Rodriguez, and R. Abd-Alhameed, "Antenna Fundamentals for Legacy Mobile Applications and Beyond", *Springer*, New York, NY, USA, **2018**.

[3]. Liao W. J, C. Y. Hsieh, B. Y. Dai and B. R. Hsiao, "Inverted-F/slot integrated dual-band four-antenna system for WLAN access points," *IEEE Antenn. Wirel. Pr.*, Vol. 14, 847–850, **2014**.

[4]. Iqbal A, O. A. Saraereh, A. W. Ahmad and S. Bashir, "Mutual coupling reduction using F-shaped stubs in UWB-MIMO antenna," *IEEE Access*, Vol. 6, 2755–2759, **2017**.

[5]. Abdullah M., S. H. Kiani and A. Iqbal, "Eight element Multiple-Input Multiple-Output (MIMO) antenna for 5G mobile applications," *IEEE Access*, Vol. 7, 134488–134495, **2019**.

[6]. Abdullah M., S. H. Kiani, L. F. Abdulrazak, A. Iqbal, M. Bashir, S. Khan, and S. Kim, "High-performance Multiple-Input Multiple-Output antenna system for 5G mobile terminals," *Electronics*, Vol. 8, 1090, **2019**.

[7]. Iqbal A., A. Basir, A. Smida, N. K. Mallat, I. Elfergani, J. Rodriguez, and S. Kim, "Electromagnetic bandgap backed millimeter-wave MIMO antenna for wearable applications," *IEEE Access*, Vol. 7, 111135–111144, **2019**.

[8]. Ahmad M. S., W. Mohyuddin, H. C. Choi, and K. W. Kim, "4 × 4 MIMO antenna design with folded ground plane for 2.4 GHz WLAN applications," *Microw. Opt. Technol. Lett.*, Vol. 60, 395– 399, **2018**.

[9]. Alibakhshikenari M, M. Khalily, B. S. Virdee, C. H. See, R. A. Abd-Alhameed, and E. Limiti, "Mutual-coupling isolation using embedded metamaterial EM bandgap decoupling slab for densely packed array antennas," *IEEE Access*, Vol. 7, 5182–51840, **2019**.

[10]. Yang L., T. Li, and S. Yan, "Highly compact MIMO antenna system for LTE/ISM applications," *Int. J. Antennas Propag.*, 714817, **2015**.

[11]. Elfergani I. and M. Sajedi, "Low profile and closely spaced four element MIMO antenna for wireless body area network," *Electronics*, Vol. 9, No. 2, 258, **2020**.

[12]. Kiem, N. K., H. N. B. Phuong, Q. N. Hieu, and D. N. Chien, "A novel metamaterial MIMO antenna with high isolation for WLAN applications," *Int. J. Antennas Propag.*, Vol. 2015, Article ID 851904, 9 pages, **2015**.

[13]. Singh H. S., G. K. Pandey, P. K. Bharti, and M. K. Meshram, "A compact dual-band diversity antenna for WLAN applications with high isolation," *Microw. Opt. Technol. Let.*, Vol. 57, 906–912, **2015**.

[14]. Likhitha T., S. Ashok Kumar, and T. Shanmuganantham, "Design of compact four port MIMO antenna using SRR ring for high isolation," *Proceedings of the International Conference on Antenna Testing and Measurement Society (ATMS 2018)*, 1–4, Pune, India, February 6–7, **2018**.

Chapter V

CHAPTER V

A HIGH-PERFORMANCE QUAD-PORT DEFECTED GROUND STRUCTURED MULTIPLE-INPUT MULTIPLE-OUTPUT ANTENNA SYSTEM FOR 5G SUB-6 GHZ RF DEVICES

The design of a miniaturized Four-element MIMO antenna system for sub-6 GHz Fifth-generation (5G) RF device applications generating different bands and high port isolation properties is presented in this work. The suggested antenna system features a single-band module comprised of four 9-shaped elements positioned perpendicularly; each antenna element in this feature covers the C-band, providing uniqueness as well as good isolation. The total dimension of the main antenna is 38x38x1.6 mm^3. A single antenna is made up of four radiating elements that are fed via a 50 Ω microstrip feeding line built on the main board. According to the results, the developed MIMO antenna radiates at multiple frequency ranges inside the assigned 5G spectrums, namely 5.0-5.2 GHz, based on impedance bandwidth criterion. It is also revealed that the antenna elements can produce pattern diversity in both frequency ranges. Furthermore, isolation of greater than 18 dB is seen between any two radiating elements. Several MIMO essential performance metrics, including as diversity gain, envelope correlation coefficient, and channel capacity are evaluated. A pattern is created, observed, and the measured and simulated results are found to agree well. Based on its performance characteristics and highest peak gain of 4.8 dBi, the proposed MIMO system can be used in 5G communication networks.

5.1 ANTENNA DESIGN

The suggested antenna concept is designed on the utilization of a MIMO antenna with four terminals as the radiating element, connected to the slotted ground. The antenna is made on an FR-4 substrate with a dielectric constant of 4.4 and a loss tangent of 0.002. The designed antenna has a centre frequency of 5.2 GHz. The length and width of the slot, as well as the length of the resonator strip, determine the antenna's bandwidth. Four rectangular-shaped slots are carved into the ground, effectively increasing the bandwidth from 5.0 to 5.2 GHz by changing the surface current distribution at higher-order modes. A single element has a compact size of 19x19 mm^{2}and the offered antenna's size for MIMO setup is 38x 38 mm^2. The offered antenna is depicted in Figure 5.1 from both the front and the back. Table 5.1 provides the proposed antenna's dimensions.

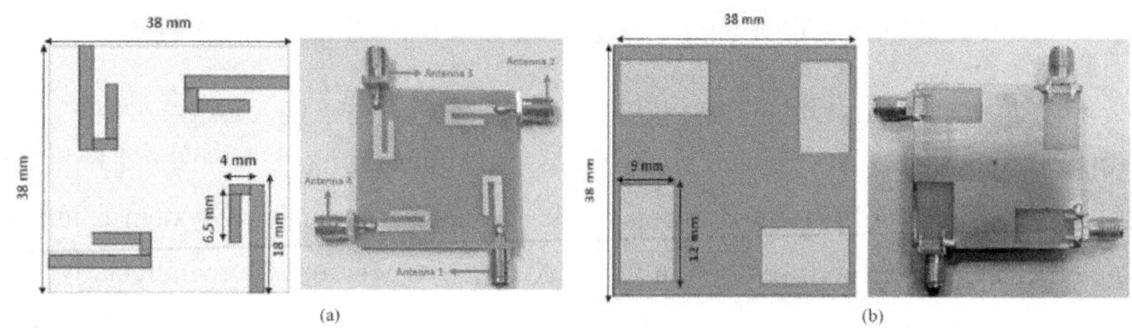

Figure 5.1. Layout of the Suggested Antenna Design (a) Front View and (b) Back View

Table 5.1. Parameters of Ultimate proposed antenna design

Parameter	Value (mm)	Parameter	Value (mm)
a	38	h	9
b	38	i	12
c	18	j	38
d	6.5	k	38
g	4	-	-

Figure 5.1 depicts the complete structure and intricate design of the quad-element antenna that has been proposed for 5G RF device applications. A primary system circuit board with a truncated ground structure and slots, is one of the PCBs of the system being demonstrated. All the PCBs which are made using an FR4 substrate is quite inexpensive. The system ground plane, which measures 38 mm x 38 mm, is printed on the backside of the main system circuit board. The quad antenna elements are made on the front side of the substrate. The antenna, which has an impedance of 50 Ω is fed through a SMA (miniature-A) connector using a rectangular microstrip line. The main challenge in building a MIMO array is mutual coupling. Mutual coupling in a set alters the element's

radiation pattern. It is proportional to the distance between the antenna elements in an arrangement. As a result, different studies used various techniques to overcome this problem. Some common approaches for reducing mutual coupling include the EBG, the DGS, the NL and the parasitic method. Some authors have used the structure decoupling method. In this work, we provide the DGS approach for reducing mutual coupling.

5.2 PARAMETRIC MODELLING

By meticulously calculating a number of parameters, the proposed antenna resonance was accomplished over multiple steps. Each parameter is essential in evaluating the antenna to resonate at the proper frequency response. The lengths of the fundamental resonator strips, as well as the length and width of the ground slot, were studied parametrically and sequentially monitored. The extensive parametric analysis of the suggested antenna system is shown in Figure 5.2. h was the first parameter to be examined. The range for h's parametric value was 9mm to 11 mm. It was found that the proposed antenna had the best resonant performance at an optimal width of 10.5 mm, and of an increase in the value led to an impedance mismatch. Following this, the factor i was examined from 12 to 14 mm, as shown in Figure 5.2 (a). The resonance frequency was found to be ahead of the required frequency at the lower beginning values, and as the parameter length increased, the resonance response shifted backward. The length at which the desired frequency response was obtained was 13.5 mm, which was the ideal length. The ground slot's length and width were crucial in adjusting and matching the antenna impedance.

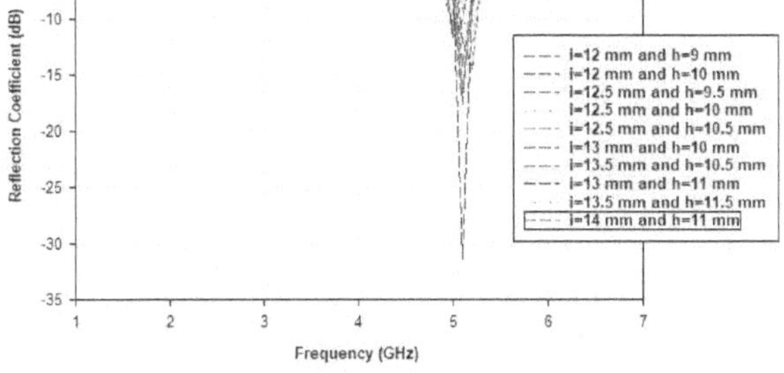

(a)

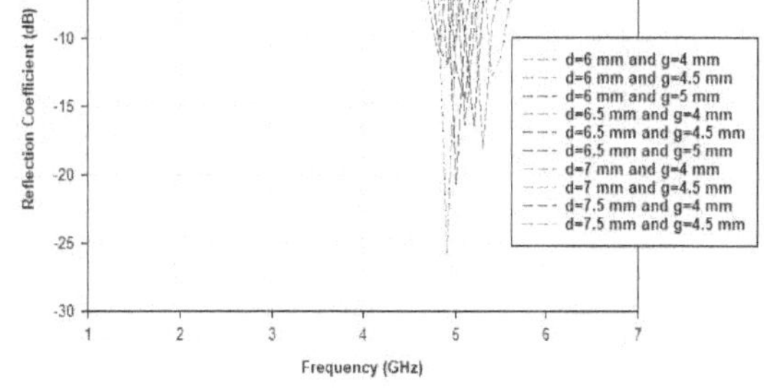

(b)

Figure 5.2. Parametric Modelling (a) with respect to i and h and (b) with respect to d and g

Figure 5.2 (b) illustrates the part that parameters d and g show in the resonance. The length of g shows a significant part in the proposed antenna resonance. The resonance shifts away from the required values with a smaller value, and 4.5 mm was the ideal length for the parameter. The resonant frequency of the suggested antenna is where parameter g plays a most crucial function, as can be shown in Figure 5.2 (b). The suggested antenna's g parameter was parametrically examined from 4 to 6.5 mm. It was noted that despite a 0.5 mm increase or decrease in value, the resonance frequency changed swiftly.

5.3 RESULTS AND DISCUSSION

The succeeding subcategories give an overview of the quad-port DGS MIMO antenna's findings. To assess its performance, certain characteristics like the S-parameter, Peak gain, efficiency, and radiation characteristics are provided.

5.3.1 Reflection and Transmission Coefficient

The antenna was tested to confirm the accuracy of the simulation. In Figure 5.3, the simulated transmission and reflection characteristics of the offered MIMO antenna are displayed. The results from simulation and measurement are very similar. S-Parameter characteristics S11, S22, S33 and S44 are presented in Figure 5.3 (a). The measured isolation that was attained as 32 dB, and the cable and measurement setup losses are to blame for the tiny discrepancy in the S-parameters. The results of Figure 5.3 (b) also shows that the isolation between two nearby antenna elements is reduced, providing low mutual coupling between the elements. The reason for the reduced isolation is that each array element has its own ground plane, which prevents current flowing from one element to the next.

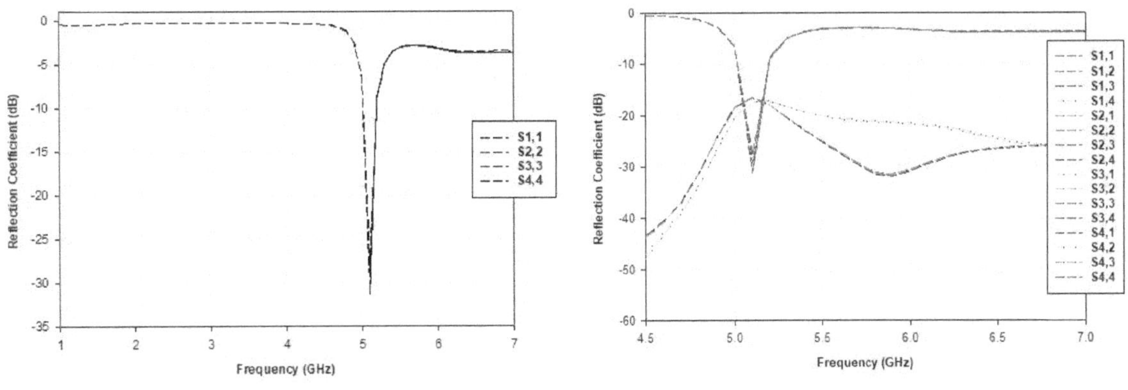

(a) (b)

Figure 5.3. S-Parameter of the given antenna (a) Reflection Coefficient with all ports and (b) Isolation Performance of the suggested antenna

5.3.2 Surface Current Distribution

The current distribution of the offered MIMO antenna for various port excitations and frequencies is shown. Current distributions caused by ports 1, 2, 3, and 4 are taken respectively and are given in Figure 5.4. It has been observed that other ports are minimally isolated when one port is excited. Surface currents can be used to identify the electric current flowing at a certain area in a conductor. To understand the impact of the coupling between the antenna sections when using this method, one of the four ports should be excited while the other three ports remain terminated with a 50 Ω load. Low isolation between MIMO components is accomplished because there is less space between the stimulated element and the left nearby ground. These changes are also evident in the S-parameter properties. The bending region experiences a strong amount of current. The ground plane's rectangular slot's edges display this constructive interface. This demonstrates that the parameter d is generating the core resonance. This parameter, which was previously described in parametric study, occurs predominantly for attaining resonance at the intended band, and any variation in length results in a dominant change in resonance for the proposed design.

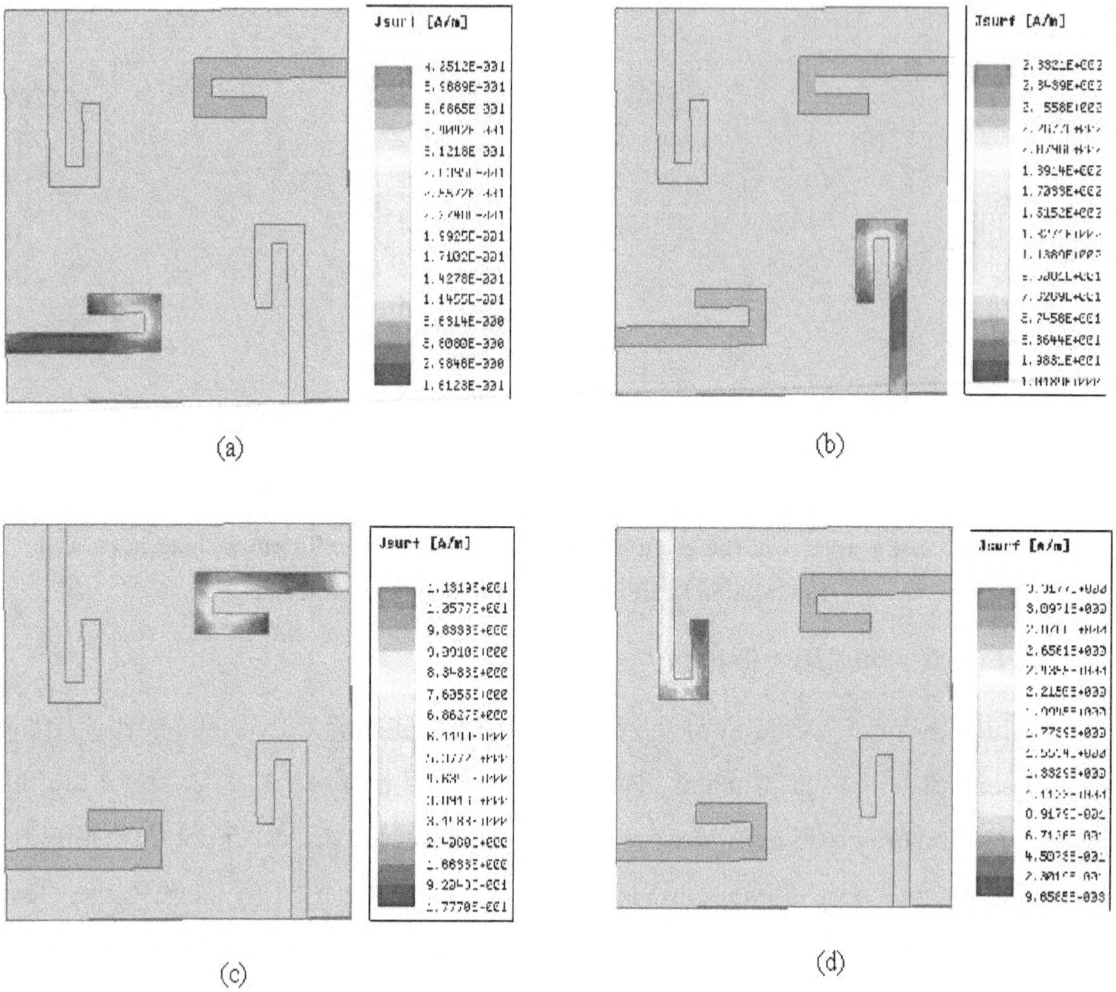

Figure 5.4. Simulated Current distribution of the MIMO antenna at 5.2 GHz, excitation of (a) Port 1, (b) Port 2, (c) Port 3 and (d) Port 4.

5.3.3 Radiation Pattern

This MIMO antenna was aimed at providing characteristics for pattern diversity. The radiation patterns of the design at theta 90 and phi 90 are shown in Figure 5.5. The results from simulation and measurement are remarkably similar. Additionally, the radiation patterns of all the radiating elements are displayed so that the diversity features attained can be clearly seen. The suggested MIMO antenna elements three-dimensional (3D) gain graphs at various resonant frequencies are shown. The predicted and measured radiation patterns of the suggested quad-port MIMO antenna in the XY, YZ, and ZX planes at port-1 and port-4 are shown in the images below.

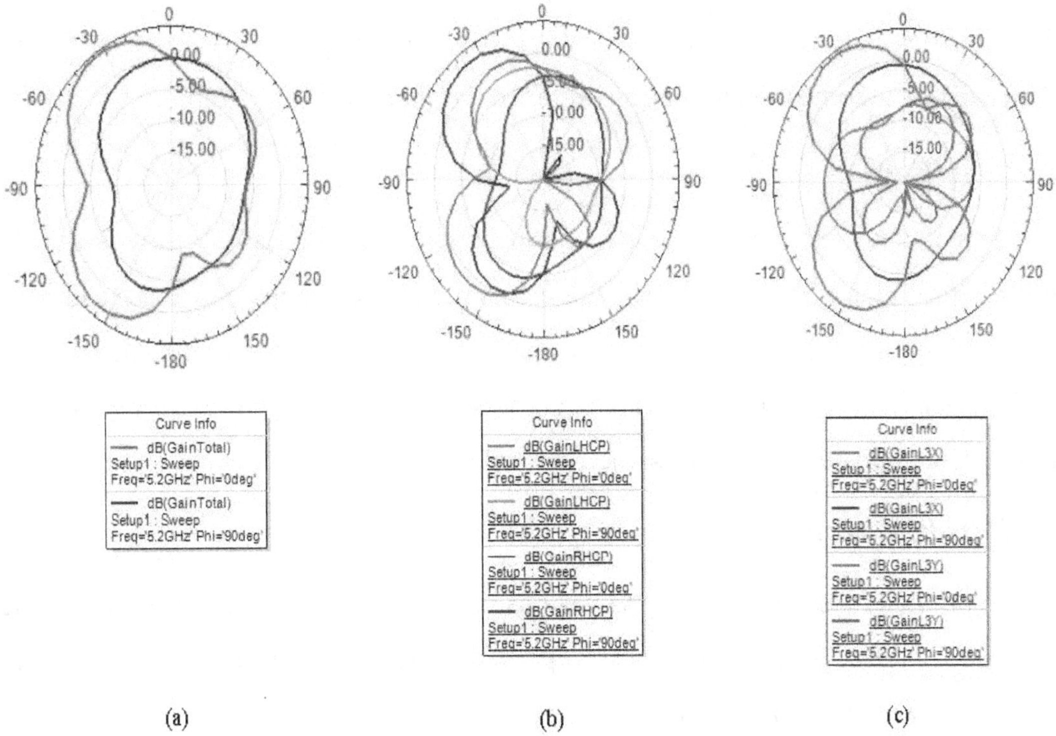

Figure 5.5. Radiation Pattern of the projected antenna (a) Both E- and H- Plane, (b) Gain LHCP and RHCP and (c) Gain L3X and L3Y

This MIMO antenna was aimed at providing characteristics for pattern diversity. The radiation patterns of the design at theta 90 and phi 90 are shown in Figure 5.5. The results from simulation and measurement are remarkably similar. Additionally, the radiation patterns of all the radiating elements are displayed so that the diversity features attained can be clearly seen.

The suggested MIMO antenna elements three-dimensional (3D) gain graphs at various resonant frequencies are shown. The predicted and measured radiation patterns of the suggested quad port MIMO antenna in the XY, YZ, and ZX planes at port-1 and port-4 are shown in the images above.

5.3.4 Performance Parameters

When examining antenna performance, the antenna characteristics are significant. Both the MIMO characteristics and the fundamental antenna performance characteristics are covered in this section [1]. The suggested antenna plots the maximum efficiency and maximum gain over frequency, allowing us to determine that the antenna's overall efficiency varies between 72 and 74%, with a maximum gain of 4.8 dBi at the stated resonance. The outcomes of simulation and measurement are comparable.

5.3.5 Group Delay

Group delay characteristics can describe the time domain behavior of MIMO antennas. It provides a list of the MIMO elements measured group delay characteristics for each of the four ports. Group Delay (m,n) describes the delay between ports m and n in this case [2]. Group delay (m,n) features are equal to (n,m) properties when m is not equal to n since MIMO antenna elements are same with balanced spacing.

5.3.6 MIMO Performance Metrics

S-parameters are used to assess various diversity factors in order to determine the effectiveness of the suggested antenna arrangement. The given metrics are used to calculate the performance of the provided antenna.

5.3.6.1 Total Active Reflection Coefficient (TARC)

Effective operational bandwidth of MIMO antenna is described by TARC characteristics. When several antennas are operating simultaneously in MIMO systems, they have an impact on one another. Therefore, it is insufficient to forecast the MIMO performance of the antenna by relying simply on ECC and DG. Together with considering factors and random signal combinations, it also includes the analysis of mutual coupling [3]. It provides a more expressive representation of MIMO performance as a result. It is described as the square root of the proportion between the total power that is reflected and the power that is incident. The expression given below is used to evaluate TARC for a dual port MIMO network. With the use of the expression in equation (5.1), where θ is the input phase, TARC may be computed from S-parameters.

$$\text{TARC} = \sqrt{\frac{(S_{11}+S_{12})^2+(S_{21}+S_{22})^2}{2}} \quad (5.1)$$

5.3.6.2 Envelope Correlation Coefficient (ECC)

It is an essential factor for characterizing the degree of correlation between antenna elements in a MIMO antenna system. Researchers frequently use ECC to evaluate how the MIMO antenna systems perform in terms of diversity. Considering that many data streams must be sent simultaneously and separately, the antenna radiators must have moderate ECC. The following formula, as demonstrated in Calculations (5.2 – 5.4), can be modified to obtain the three ECC readings in this instance by substituting the relevant terms.

$$\rho_{e12}=\frac{|S_{11}S_{12}+S_{12}S_{22}+S_{13}\,S_{32}+S_{14}S_{42}|^2}{\left(1-(|S_{11}|^2+|S_{12}|^2\,)+(|S_{13}|^2+\,|S_{14}|^2)\right)^2} \tag{5.2}$$

$$\rho_{e13}=\frac{|S_{11}S_{13}+S_{12}S_{23}+S_{13}\,S_{33}+S_{14}S_{43}|^2}{\left(1-(|S_{11}|^2+|S_{12}|^2\,)+(|S_{13}|^2+\,|S_{14}|^2)\right)^2} \tag{5.3}$$

$$\rho_{e14}=\frac{|S_{11}S_{14}+S_{12}S_{24}+S_{13}\,S_{34}+S_{14}S_{44}|^2}{\left(1-(|S_{11}|^2+|S_{12}|^2\,)+(|S_{13}|^2+\,|S_{14}|^2)\right)^2} \tag{5.4}$$

5.3.6.3 Diversity Gain

Another helpful criterion for assessing the MIMO performance of the proposed antenna is diversity gain (DG). It is calculated by using the equations (5.5 – 5.6).

$$ADG = 10\sqrt{1-|ECC|}^{2} \tag{5.5}$$

$$EDG = \eta_{total}\; x\; ADG \tag{5.6}$$

ECC (m, n) and DG (m, n) characteristics are same with ECC (n, m) and DG (n, m) when m not equal to n since MIMO antenna elements are similar with even spacing. ECC values in the operational bands are observed to be less than 0.1, and DG is extremely close to 10, indicating that the proposed MIMO antenna has successful diversity characteristics. The SNR after implementing the MIMO system is effectively increased as measured by the ADG. The acceptable maximum ADG level is 10 dB. The EDG is another practical DG variant that accounts for antenna radiation losses. The ECC values calculated above are used to calculate the ADG and EDG using formulas (5.5) and (5.6). For a MIMO system to work effectively, both the ADG and EDG must be higher than 7.5 dB.

5.3.6.4 Mean Effective Gain

Another helpful measure for estimating MIMO performance is MEG. The MIMO antenna is built from identical antennas, hence the gain provided by individual antenna parts should not vary significantly. For MEG, 0 dB is the optimal theoretical limit. However, a ± 3 dB deviation is acceptable. It is the ratio between the power received by an isotropic antenna and by a diversity antenna. Equation (5.7) describes how to calculate MEG_i of MIMO element/port-i with 'n' number of elements/ports for MIMO systems. Since MEG_i is -3 dB at all frequencies, the suggested MIMO antenna performs well in terms of diversity.

$$MEG_i = 0.5[1 - \sum_{j=1}^{n} |S_{ij}|^2] \quad (5.7)$$

5.4 COMPARISON

The proposed MIMO antenna characteristics from the state-of-the-art literature are compared with other relevant studies in Table 5.2 in relation to sizes, operating band, number of MIMO elements, gain, radiation efficiency, isolation, etc.

Table 5.2. Performance Comparison from other MIMO Antennas

Ref. No	Dimension (mm³)	ε_r	No. of elements	Operating frequency (GHz)	Peak gain (dBi)	Efficiency (%)	Inter port Isolation (dB)
[4]	120x65x1.6	4.4	4	3.3-5.0	4.71	NR	>18
[5]	145x75x0.8	4.4	8	3.5&5.0	NR	>35	NR
[6]	150x70x0.8	4.4	8	3.5&5.0	4.4&5. 5	>61	>20
[7]	20x24x24	4.4	4	3.0-6.75	4.97	>80	>15
[8]	19x5x0.8	4.4	2	3.3-3.6	5.9	>76	>15
This Work	**38x38x1.6**	**4.4**	**4**	**5.0-5.2**	**4.8**	**>74**	**>32**

5.5 CONCLUSION

The proposed antenna is intended for 5G sub-6 GHz applications and is described in this paper. It works in the region from 5.0 to 5.2 GHz and covers C-band application. With suitable antenna element configuration and optimised tiny slots in grounding, element-to-element isolation of more than 18 dB is achieved. The design's radiation performance is evaluated using efficiency, gain, and pattern results. A correctly tuned design achieves stable and bore sight patterns. To assess the MIMO performance of the offered architecture in a free space environment, the diversity metrics are employed. It is possible to get low ECC, high DG, decent TARC, low CCL, and acceptable MEG readings. These are related to measured ones, and they fit perfectly. The suggested antenna is made of low-cost dielectric FR4 material and has a dimension of 38x38x1.6 mm^3. Table 5.2 compares the proposed study to the other works identified in the literature. The proposed antenna features a new configuration, higher isolation, lower CCL values, and higher efficiency values as compared to the ones stated, and the structure is developed exclusively for 5G sub-6 GHz applications.

References

[1]. Jing Cai, Guiting Dong, Chuanba Zhang, Huiyang Li, Xiaojing Shi and Gui Liu," Design and characterization of a minimized planar dual-band MIMO circular ring antenna with high efficiency", *Microwave and Optical technology letters*, 619-652, **2022**.

[2]. Jaiverdhan, M.M.Sharma and R.P.Yadhav," Broadband circularly polarized compact MIMO slot antenna based on strip and stubs for UWB applications", *Electromagnetics*, 41, 185-195, **2021**.

[3]. He Peng, Ruixing Zhi, Qichao Yang, Jing Cai, Yi Wan and Gui Liu," Design of a MIMO antenna with high gain and enhanced isolation for WLAN applications", *Electronics*, 10,1659, July **2021**.

[4]. Zhang T, Chen L, Moghaddam, S. M Zaman, A. U and Yang J," Ultra-wideband linearly polarised planar bowtie array antenna with feeding network using dielectric-based inverted microstrip gap waveguide", *IET Microwaves, Antennas & Propagation*, **2020**.

[5]. Das G, Sahu N.K, Sharma, Gangwar R.K, Sharawi and M.S," FSS-Based Spatially Decoupled Back-to-Back Four-Port MIMO DRA with Multidirectional Pattern Diversity", *IEEE Antennas Wirel. Propag. Lett.*, 18, 1552–1556, **2019**.

[6]. Jaiswal R.K, Kumari K, Ojha A.K and Srivastava K.V, "Five-port MIMO antenna for n79-5G band with improved isolation by diversity and decoupling techniques", *J. Electromagn. Waves Appl.*, 35, 1–15, **2021**.

[7]. Dwivedi A.K, Sharma A, Singh A.K and Singh V," Design of dual band four port circularly polarized MIMO DRA for WLAN/WiMAX applications", *J. Electromagn. Waves Appl*, 34, 1990–2009, **2020**.

[8]. Sarkar D and Srivastava K.V, "Compact four-element SRR-loaded dual-band MIMO antenna for WLAN/WiMAX/WiFi/4G-LTE and 5G applications", *Electr. Lett.*, 53, 1623–1624, **2017**.

[9]. Rayirathil K. Athira Mohan and Kanagasabapathi G. Padmasine," A Slotted Compact Four-Port Truncated Ground Structured MIMO Antenna for Sub-6 3.4 GHz 5G Application", *Progress in Electromagnetics Research C*, 122, 267–277, **2022**.

Chapter VI

CHAPTER VI

DESIGN AND ANALYSIS OF A NOVEL-SHAPED HEXA-BAND FREQUENCY RECONFIGURABLE ANTENNA INTEGRATED WITH PIN DIODE FOR S-BAND AND 5G SUB-6 GHZ APPLICATIONS

The work presents a novel shaped hexa-band frequency reconfigurable antenna. Depending upon the switching states, it can take place with three Switching states. For the three switching, two single band modes (3.5 and 5.4 GHz) and two dual band modes (2.29, 4.21, 2.6 and 5.2 GHz). The desired operating characteristics were accomplished using a number of switching states. The intended antenna performs in a dual band mode when all of the switches are switched ON, and it operates in a single band mode when all of the switches are turned OFF. It functions in a dual band mode when two switches are in the ON position and one switch is in the OFF position. When both switches were OFF and one switch was ON, the antenna performed on a single resonant band. In the simulation platform, lumped constituents are used to generate tunable capacitance, which is liable for frequency reconfigurability. The fabricated antenna's evaluated tuning capability covers from 2.29 GHz to 5.2 GHz. For the different operating conditions of the described hexa-band antenna, remarkable efficiencies, suitable radiation patterns, appropriate gain, enhanced impedance matching, and directivity values were attained. Ansys HFSS is used to examine the far field and scattering properties. The developed antenna may very well be seamlessly implemented into modern communications systems due to its compact and inexpensive architecture. To evaluate the simulation results, a prototype of the proposed antenna is fabricated and analysed using PIN diode switches. The measured and simulated findings for the proposed reconfigurable antenna are in excellent agreement.

6.1 PARAMETRIC STUDY OF THE DESIGNED ANTENNA

The basic geometry, conceptual design, and switching mechanisms of the suggested hexa-band frequency reconfigurable monopole antenna are discussed and summarized in this section. The proposed antenna is altered in the simulation environment through lumped element switches to achieve two single band and two dual band modes. Furthermore, to achieve reconfigurability, PIN diodes are incorporated in the measuring setup. To enhance efficiency and achieve satisfactory far field radiation patterns, a shortened metallic ground plane is employed on the rear. Figure 6.1 illustrates the structural dimensions and layout of the designed hexa-band frequency reconfigurable antenna for UMTS, Wi-Fi, WiMAX, WLAN and 5G applications. The proposed antenna's radiating element is printed on a lossy FR-4 substrate with a relative permittivity of 4.4 and a tangent loss of 0.02, which is supported by a truncated metallic ground surface. Because the FR-4 substrate is widely available, the antenna design is more inexpensive and viable. The truncated metallic ground surface has the characteristic of providing the best gain, efficiency, and directivity. For antenna excitation, a 50 Ω micro strip line with a width of 3 mm is employed. The antenna is excited using the waveguide port attached to the feed line. As demonstrated in Figure 6.1, four 1 mm wide slots in the radiating structure are designated for the implementation of lumped element switches. The suggested monopole antenna has a volume of 33x16x1.6 mm^3. Table 6.1 illustrates the proposed structure's specific dimensions.

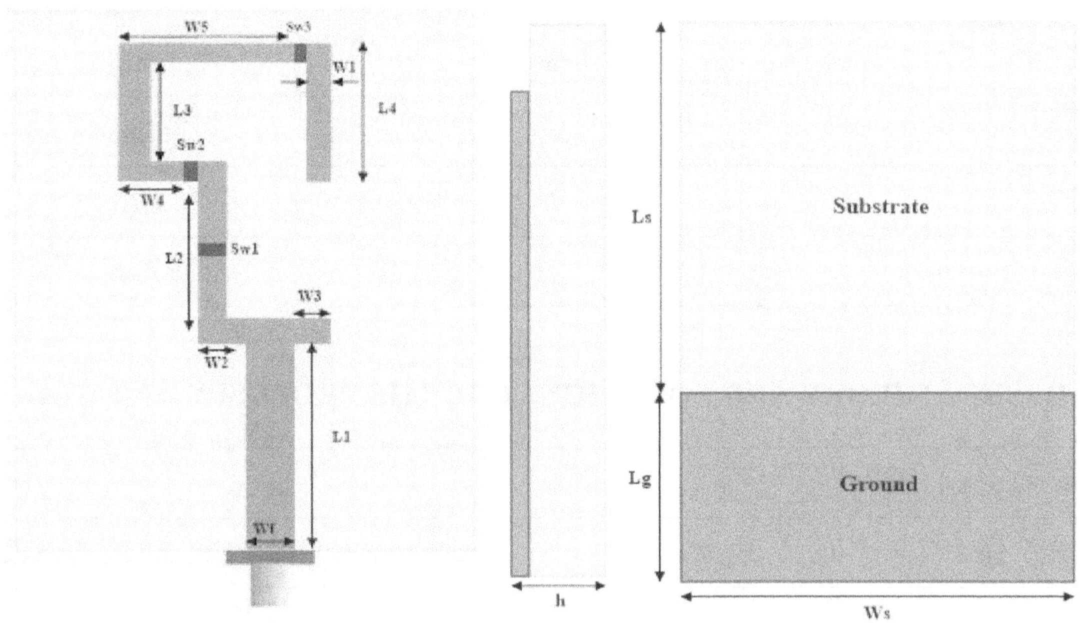

(a) **(b)**

Figure 6.1. Layout of the designed Hexa-Band Frequency Reconfigurable Antenna with three PIN diodes: (a) Front view (b) Side view (c) Rear view

Table 6.1. Dimensions of the Proposed Design

Sl. No	Dimensions	Values
1	L_1	15 mm
2	L_2	9.1 mm
3	L_3	4.5 mm
4	L_F	6.8 mm
5	L_S	33 mm
6	L_G	7.5 mm
7	W_1	2 mm
8	W_2	2.5 mm
9	W_3	3 mm
10	W_4	2 mm
11	W_5	6.3 mm
12	W_F	3 mm
13	W_S	16 mm
14	H	1.6 mm

6.2 DESIGN THEORY AND EQUATIONS FOR THE PROPOSED ANTENNA

$$S_{2.29} = \lambda_{2.29}/4 \tag{6.1}$$

$$S_{2.6} = \lambda_{2.6}/4 \tag{6.2}$$

$$S_{3.5} = \lambda_{3.5}/4 \tag{6.3}$$

$$S_{4.21} = \lambda_{4.21}/4 \tag{6.4}$$

$$S_{5.2} = \lambda_{5.2}/4 \tag{6.5}$$

$$S_{5.4} = \lambda_{5.4}/4 \tag{6.6}$$

For the interconnection of the switches within the proposed design, 1 mm slots in the radiating patch are designated. The transmission line model principle is used to evaluate the resonant lengths of the proposed antenna, given in equations 6.1 – 6.6, which are $S_{2.29}$, $S_{2.6}$, $S_{3.5}$, $S_{4.21}$, $S_{5.2}$, and $S_{5.4}$. Changing the switching states results in the resonant frequencies modes. The directed wavelength and resonant lengths are referred to as [1-4]:

The directed wavelength and antenna quality were also maximized for adequate radiation efficiency, which can be calculated using the equation (6.7). The surface current distribution for a particular frequency mode is being used to calculate the appropriate resonant length, which is expressed in terms of directed wavelengths, l_g, as given [5-7]:

$$Lfr = \frac{\lambda g}{4} = \frac{C}{4fr\sqrt{\varepsilon e}} \tag{6.7}$$

Where, C is the velocity of light and f_r is the resonance frequency and ε_e is the effective permittivity. The effective permittivity can be expressed in the given equation (6.8)

$$\varepsilon e = \frac{\varepsilon r+1}{2} + \frac{\varepsilon r-1}{2}(1+12h/w)^{-0.5} \tag{6.8}$$

Where, W is the width of radiating element of the antenna, h is the height of substrate, and ε_r is the relative permittivity. When the antenna is excited at the exact point, the efficiency increases, resulting in a decrease in the reflection coefficient, which is described in a mathematical model as:

$$|\Gamma| = \frac{Za - Zc}{Za + Zc} \tag{6.9}$$

Where, Z_a is the antenna impedance and Z_c is the characteristic impedance of the feed line. The reflection coefficient (S11) and VSWR are multiple antenna parameters that are linked. The definition of VSWR is

$$VSWR = \frac{1+|\Gamma|}{1-|\Gamma|} \tag{6.10}$$

In the desired frequency range, the antenna has a relatively low reflection coefficient when exactly suited. The radiation efficiency associates the proposed antenna's directivity and gain. The gain is commonly expressed in decibels (dB) and is stated as follows [8-10]:

$$G(dB) = 10\log 10(\eta radD) \qquad (6.11)$$

Where, G is the gain and D is the directivity of the proposed antenna.

6.2.1 Switching Techniques

PIN diodes function as variable resistors at RF frequencies. PIN diodes, on the other side, have slightly more complicated circuitry for the ON and OFF states. In the equivalent circuit, PIN diodes have an inductance (L_b) in both ON and OFF states. The low resistance (R_{bS}) was included in the ON state which is forward biased, while the equivalent PIN was included in the OFF state and is reverse biased. As shown in Figure, a diode circuit is made up of a parallel combination of resistance (R_{bP}) and capacitance (C_c). As possible, the PIN diodes are structured as two lumped RLC boundary conditions, as shown in the Figure 6.2. The suggested antenna's reconfigurable features have been designed to simulate only by using resistors. Users did not consider capacitance and inductance values in the simulation because the switch is on the ON-state in both the RLC lumped model and the PIN diode model provides a short circuit characteristic, enabling current to flow along the structure's radiating path. In the OFF state, however, both models exhibit an open circuit phenomenon, impeding the flow of current. The equivalent circuit model for both the ON and OFF states of the PIN diode model has already been discussed [11-15].

According to the research, the ON state in the PIN diode model is recognized as a series RL component with small values that performs as a short circuit, allowing current to flow along the radiator. In contrast to this, the OFF state of a PIN diode is simulated as a parallel RLC component with such values that it demonstrates an open circuit behaviour in which current cannot flow along the radiator. To simplify our model, designers focused on open circuit and short circuit actions, and the switch was discovered as an RLC lumped element model with only resistor values. A resistor R_{bS} with a limited value of 1Ω acts as a short circuit and enables normal current flow. A resistance R_{bP} of 1 MΩ, on the other hand, demonstrates open circuit behaviour and stops the path for current to flow along the radiating structure [16-19].

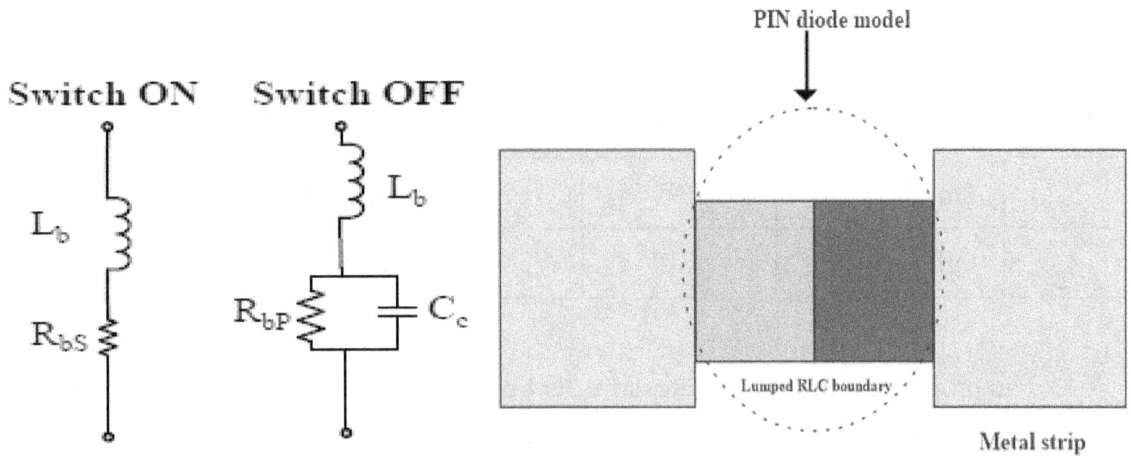

(a) **(b)**

Figure 6.2. RF PIN Diode Configuration (a) Equivalent circuit models for the ON and OFF states of the switch (b) PIN Diode model.

6.2.2 Simulated and Measured Results

The radiating structure is designed, modelled, and examined using Ansys HFSS to investigate the performance of the hexa-band antenna. A waveguide port is often used as an excitation source for the proposed antennas. The VSWR, gain, directivity, far field pattern, reflection coefficient, and surface electric-field are measured in Ansys HFSS using the transient solver with open-add-space boundary conditions.

Table 6.2. Tuning states of the Hexa-band frequency antenna

Sl. No	Sw1	Sw2	Sw3	Frequency states
1	On	On	On	Dual band (2.29 GHz and 4.21 GHz)
2	On	On	Off	Dual band (2.6 GHz and 5.2 GHz)
3.	On	Off	Off	Single band (3.5 GHz)
4.	Off	Off	Off	Single band (5.4 GHz)

The proposed antenna operates in a dual band frequency mode with a return loss of -32.9435 and -17.2412 dB at 2.29 and 4.21 GHz, respectively, while all three switches (Swl, Sw2, and Sw3) are in the ON state having a gain of 1.85 dBi and 2.75 dBi respectively. The antenna operates in dual band mode (2.6 and 5.2 GHz) with a return loss of -45.8179 and -29.3527 dB, having a gain of 2.21 dBi and 3.46 dBi respectively, when the two switches (Swl and Sw2) are ON and Sw3 is OFF. The antenna operates in a single band mode at 3.5 GHz with a return loss of -24.3420, if one switch (Swl) is switched ON and the other two switches (Sw2 and Sw3) are turned OFF with the gain of 3.57 dBi. The designed antenna operates in a single band mode at 5.4 GHz with a return loss of -17.1955 dB when all the switches are turned OFF having the gain of 4.42 dBi. Multiple processes and techniques of the antenna are accomplished using three lumped elements (RLC) switches, as indicated in Table 6.2. The lumped element switch was chosen because it can be simply modelled in Ansys HFSS as a 1 Ω and 1MΩ resistor, respectively, to describe the switch-ON and switch-OFF states. Figure 6.3 depicts the simulated and measured reflection coefficients of the antenna.

Table 6.3. Detailed summary of the simulated results of the Hexa-band frequency antenna

Parameters	Sw(1,2,3) On	Sw(1,2,3) Off	Sw(1,2) On, 3 Off	Sw1 On, (2,3) Off
Frequencies (GHz)	2.29 GHz and 4.21 GHz	5.4 GHz	2.6 GHz and 5.2 GHz	3.5 GHz
Gain (dBi)	1.85 and 2.75	3.26	2.21 and 3.46	2.38
Return loss (dB)	-32.9435 and -17.2412	-17.1955	-45.8179 and -29.3527	-24.3420
Directivity (dBi)	2.72 and 3.64	4.42	2.64 and 4.85	3.57
Bandwidth (%)	9.3 and 11.7	43.86	14.31 and 14.58	22.42
Efficiencies (%)	91.54 and 89.53	87.08	97.52 and 84.21	96.45

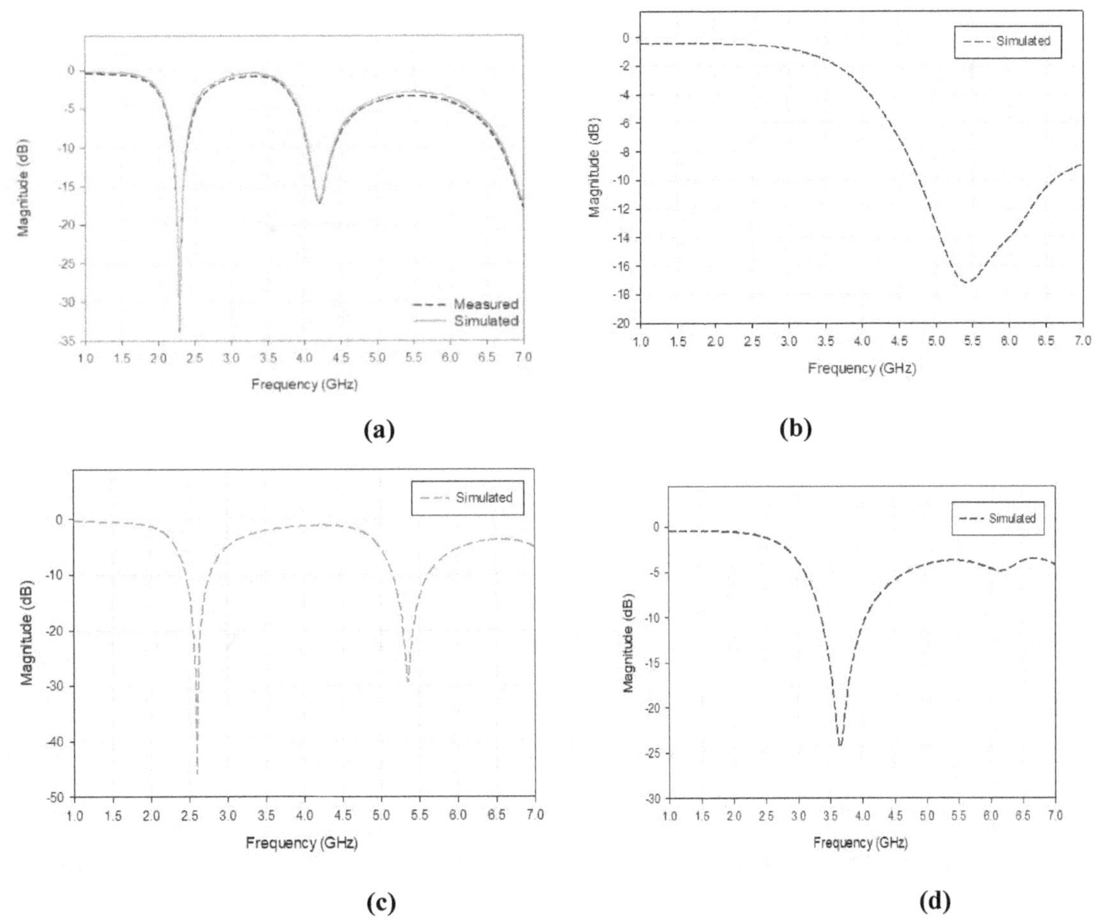

Figure 6.3. Simulated and Measured Reflection Coefficients of the operation states (a) Dual band (2.29 GHz and 4.21 GHz), (b) Single band (5.4 GHz), (c) Dual band (2.6 GHz and 5.2 GHz), (d) Single band (3.5 GHz).

PIN diodes (Skyworks SMP 1345-079LF) with low capacitance and strong result up to 6 GHz were specified for the experimental setup. When the diode is forward biased, it functions as a series resistor with a value of 1.5, whereas when reverse biased, it operates as a series capacitor with a value of 0.17 pF. It's important to remember that the constructed antenna's feeding line is totally separated from the DC path. Although capacitors can stop DC while allowing RF signals, and RF chokes can restrict RF while passing DC, an integration of the two issues should suffice. For RF choking, 125 nH inductors are required, and 470 pF capacitors are often used to short the RF signals released from the choking inductor. Two Alkaline AA batteries do provide biased voltage of 3V and the biased current of roughly 10 mA. An Agilent N5242A PNA-X Series Vector Network Analyzer is used to test the reconfigurable monopole antenna's reflection coefficient (S11). The VSWR of the designed antenna is less than 1.3 for all the frequency bands, which proves that the antenna is optimally matched. VSWR at 2.29, 4.21, 2.6, 5.2, 3.5, and 5.4 GHz is as given in Figure 6.4 and the Fabricated prototype is given in Figure 6.5.

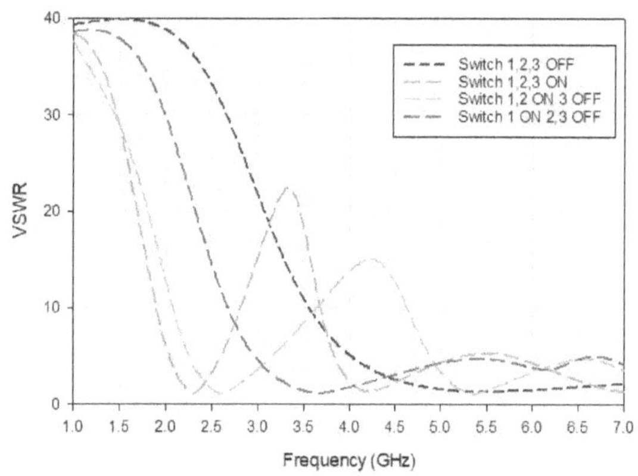

Figure 6.4. VSWR of the proposed antenna.

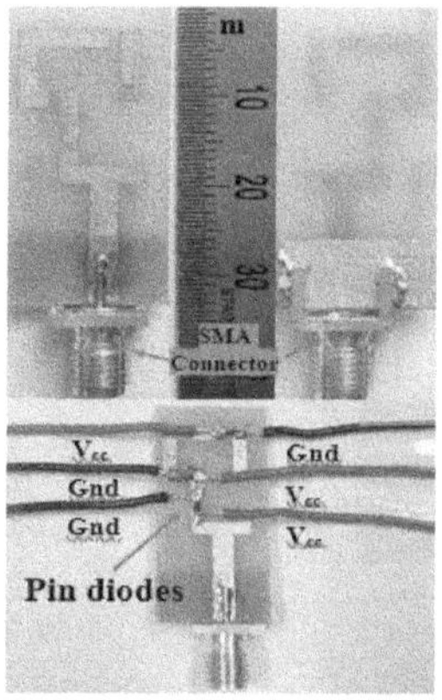

Figure 6.5. Fabricated Prototype

An azimuth turntable is used to mount the manufactured antenna on the elevation. Because the antenna is linearly polarized, the H- and E-plane radiation patterns are examined to get insight into the antenna's radiation characteristics in the two orthogonal planes. It's worthy to note that the designed hexa-band antenna radiates in an Omni direction and has dominating patterns in the horizontal H-plane throughout all six frequency bands.

The antenna is designed in a single band mode with a peak gain of 3.26 dBi at 5.4 GHz by altering the modes of the switch. The suggested antenna operates in a dual band mode with peak gain values of 2.21 dBi and 3.46 dBi at 2.6 GHz and 5.2 GHz frequency bands, respectively, while SW1 and SW2 switches are ON and SW3 switch is OFF. The antenna operates in an only one resonant mode with a peak gain value of 2.38 dBi at 3.5 GHz when SW2 and SW3 are turned OFF and SW1 is turned ON. The designed antenna's determined H-plane exhibits an omnidirectional radiation pattern, whereas the E-plane has a substantially "figure of 8" form in all six resonant bands, as shown. The 3D far-field radiation patterns for the respective frequency bands are provided in Figure 6.6 for further understanding.

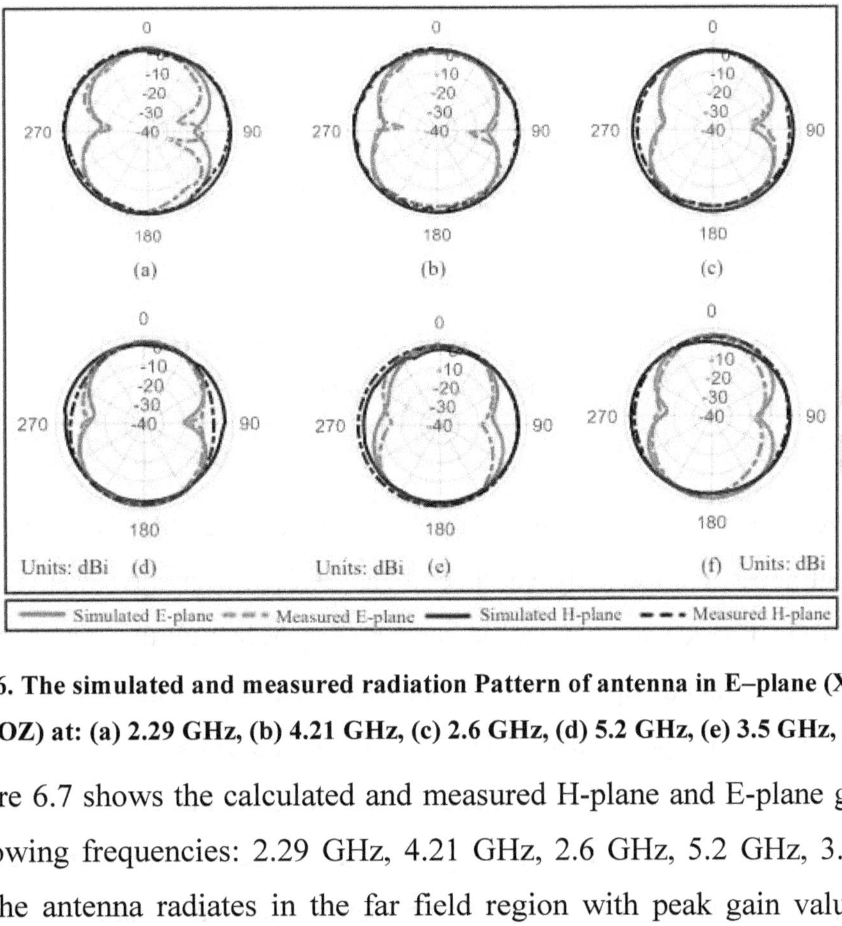

Figure 6.6. The simulated and measured radiation Pattern of antenna in E–plane (XOY) and H–plane (XOZ) at: (a) 2.29 GHz, (b) 4.21 GHz, (c) 2.6 GHz, (d) 5.2 GHz, (e) 3.5 GHz, (f) 5.4 GHz.

Figure 6.7 shows the calculated and measured H-plane and E-plane gain patterns for the following frequencies: 2.29 GHz, 4.21 GHz, 2.6 GHz, 5.2 GHz, 3.5 GHz, and 5.4 GHz. The antenna radiates in the far field region with peak gain values of every frequency are given below. Because of the matched feeding mechanism used throughout the design at the desired alignment, the proposed antenna radiates efficiently. The radiation efficiency of the designed antenna is 91.54%, 89.53 %, 87.08%, 97.52%, 84.21% and 96.45% at 2.29 GHz, 4.21 GHz, 5.4 GHz, 2.6 GHz, 5.2 GHz and 3.5 GHz frequency bands respectively. The table following summarizes the entire performance of the proposed hexa-band antenna. The suggested hexa-band antenna is analysed to the literature in the table below, which shows that the designed antenna is considerably smaller, has the most operating bands, has the highest gain, and provides the most bandwidth when related to the other works.

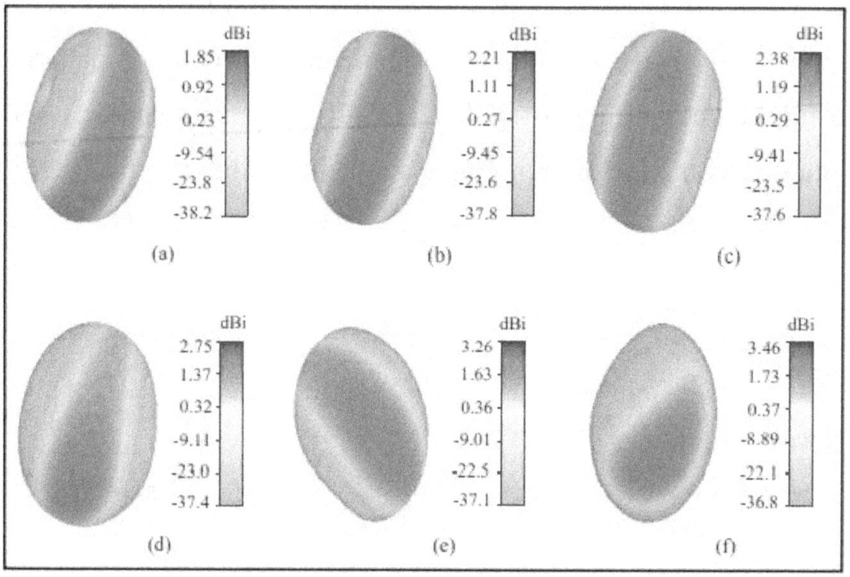

Figure 6.7. The 3D far-field gain pattern at: (a) 2.29 GHz. (b) 4.21 GHz. (c) 2.6 GHz. (d) 5.2 GHz. (e) 3.5 GHz. (f) 5.4 GHz.

6.3 CONCLUSION

In this work, a Novel hexa-band frequency reconfigurable monopole antenna has been designed and experimentally validated. The designed antenna was reconfigured to operate in two dual bands and two single band modes. When all the switches were in ON state, the proposed antenna operated in a dual band mode and when all the switches were in OFF state, the antenna performed in a single band mode. If two were in ON and one was in OFF state, it is operated in a dual band and when two switches were in OFF and one was in ON state, the antenna operated at a single resonant band. The outstanding efficiencies, satisfactory radiation patterns, acceptable gain, better impedance matching and directivity values were achieved for the different operating conditions of the proposed hexa-band antenna. The designed antenna has many advantages due to its characteristics such as light weight, compact size, low-cost, easy to fabricate and also simple in integration. The designed antenna can be intended for use in the satellite communications, military purposes and in modern communication devices such as laptops and tablets. Moreover, the measurement was conducted for the reflection coefficient and radiation patterns using PIN diodes in the fabricated prototype. The measured results of the proposed antenna were in a good agreement with the simulated results.

References

[1]. Osman M N, Rahim M K A, Gardner P, Hamid M R, Yusoff M F M and Majid H A," An electronically reconfigurable patch antenna design for polarization diversity with fixed resonant frequency", Radio Eng., 24(1):45–53, **2015**.

[2]. Saraswat R K and Kumar M," A frequency band reconfigurable UWB antenna for high gain applications", Prog Electromagn Res., 64:29–45, **2015**.

[3]. Yeom I, Choi J, Kwoun S, Lee B and Jung C, "Analysis of RF front-end performance of reconfigurable antennas with RF switches in the far field", Int J Antennas Propag.,**2014**.

[4]. Ullah S, Hayat S, Umar A, Ali U, Tahir F A and Flint J A, "Design, fabrication and measurement of triple band frequency reconfigurable antennas for portable wireless communications", AEU-Int J Electron Commun., 81:236–242, **2017**.

[5]. Shah I A, Hayat S, Khan I, Alam I, Ullah S and Afridi A," A compact, tri-band and 9-shape reconfigurable antenna for WiFi, WiMAX and WLAN applications", Int J Wirel Microw Technol., 5:45–53, **2017**.

[6]. Anand S and Gloria J J, "RF MEMS based reconfigurable rectangular slotted self-similar antenna", Circuits and Systems, 7(06): 859, May 4, **2016**.

[7]. Saleh A A and Abdullah A S," A novel design of patch antenna loaded with complementary split-ring resonator and L-shape slot for (WiMAX/WLAN) applications", Int J Wirel Microw Technol IJWMT, 4(3):16–25, **2014**.

[8]. Khan M F, Shah S A and Ullah S, "Dual-band frequency reconfigurable microstrip patch antenna on wearable substrate for Wi-Fi and Wi-MAX applications", Technical Journal, University of Engineering and Technology, Taxila, Pakistan, 22:35–40, **2017**.

[9]. Hayat S, Shah I A, Khan I, Alam I, Ullah S and Basir A," Design of tetra-band frequency reconfigurable antenna for portable wireless applications, "Intelligent Systems Engineering (ICISE), IEEE, (pp. 10–13), Jan 15, **2015**.

[10]. Hamid M R, Gardner P, Hall P S and Ghanem F," Vivaldi antenna with integrated switchable band pass resonator", IEEE transactions on antennas and propagation, 59(11):4008–4015, **2017**.

[11]. Iqbal A, Ullah S, Naeem U, Basir A and Ali U," Design, fabrication and measurement of a compact, frequency reconfigurable, modified T-shape planar antenna for portable applications", Journal of Electrical Engineering and Technology, 12(4):1611–1618, **2017**.

[12]. Ahyat E N, Kamarudin M R, Rahman T A, Jamlos M F, Hamid M R, Jamaluddin M H and Ramli N H, "Frequency and pattern reconfigurable dipole Yagi antenna", Microwave and Optical Technology Letters, 55(2):447–450, **2013**.

[13]. Li Y and Li W, "A circular slot antenna with wide tunable and reconfigurable frequency rejection characteristic using capacitance loaded split-ring resonator for UWB applications", Wireless personal communications, 78(1):137–149, **2014**.

[14]. Shah S A, Khan M F, Ullah S, Basir A, Ali U and Naeem U," Design and measurement of planar monopole antennas for multi-band wireless applications", IETE Journal of Research, 63(2):194–204, **2017**.

[15]. Ali T, Khaleeq M M and Biradar R C," A multiband reconfigurable slot antenna for wireless applications", AEU-International Journal of Electronics and Communications, 84:273–280, **2018**.

[16]. Ullah S, Ahmad S, Khan B A and Flint J A," A multi-band switchable antenna for Wi-Fi, 3G Advanced, WiMAX, and WLAN wireless applications", International Journal of Microwave and Wireless Technologies, 1–7, **2018**.

[17]. Sumit, "Reconfigurable fractal tree antenna for multifrequency applications using PIN diode", International Journal of Engineering Research and Technology (IJERT), Vol. 3 Issue 4, April **2014**.

[18]. Li Y, Li W and Mittra R, "A CPW-fed wide-slot antenna with reconfigurable notch bands for UWB and multiband communication applications", Microwave and Optical Technology Letters, 55(11):2777–2782, **2013**.

[19]. Li Y, Li W and Mittra R," A compact CPW fed circular slot antenna with reconfigurable dual band notch characteristics for UWB communication applications", Microwave and Optical Technology Letters,56(2):465–468, **2014**.

Chapter VII

CHAPTER VII

CONCLUSION AND FUTURE PERSPECTIVE

7.1 THESIS HIGHLIGHTS

This chapter brings the thesis to a close by presenting the conclusions drawn from the outcome of various MIMO antennas for 5G Applications. The main objective of the thesis was to develop antennas with good impedance matching and to reduce mutual coupling having greater antenna efficiencies. Another objective was to design compact MIMO antenna structures suitable for small mobile terminals. Since multiple antenna systems have large size, design of compact antenna was a demanding task. The experimental and simulation investigations on the possibility of resonance frequencies are extensively explained.

Chapter one provides an introduction on Reconfigurable antenna research, state of the art antenna technologies, MIMO antenna research, Applications of MIMO, facilities and techniques for antenna fabrications and measurements. Some over view on the significant work in related areas, provided in chapter two gives real picture about the importance of the present work. The literature review clearly indicates that the antennas developed are novel concepts. Chapter three describes a compact and closely spaced Four-port MIMO antenna for 5G applications. Chapter four discusses a slotted compact Four-port truncated ground structured MIMO antenna for Sub-6 3.4 GHz for 5G applications. A High-performance Quad-port defected ground-structured MIMO antenna system for 5G Sub-6 GHz RF devices is presented in chapter five. Chapter six introduces the Design and analysis of a Novel-shaped Hexa-Band Frequency reconfigurable antenna integrated with PIN diode for S-band and 5G Sub-6 GHz applications.

7.1.1 Inferences from the investigation on Four-port MIMO antenna

A compact and closely spaced four port MIMO antenna for 5G applications is discussed in this chapter. The important inferences from this study are as follows:

- The presented Four port MIMO antenna is providing a high impedance matching whose isolation is about -27 dB.

- The offered MIMO antenna deliver high radiation efficiency with typical peak realised antenna gain.
- A compact high vólume and adequate bandwidth with good accomplished gain making the antenna a very strong for 5G application.
- The recommended antenna works effectively for applications in both mobile and satellite communications as well as in medical applications.

7.1.2 Inferences from the investigation on substrate modified antenna

A Slotted Compact Four-port Truncated Ground Structured MIMO antenna for Sub-6 3.4 GHz 5G Application is demonstrated in this chapter.

- The proposed MIMO antenna having broad frequency bandwidth, excellent gain, decreased inter-element gap and effective isolation is clearly novel.
- The four-element MIMO system has been situated orthogonally to the others while maintaining a small size and a good result.
- The given substrate altered antenna provides a high impedance matching whose isolation is about -28 dB.
- The recommended antenna provides effectively for various mobile applications.

7.1.3 Inferences from the High-performance Quad-port DGS MIMO antenna for 5G Sub-6 GHz RF devices

- The given miniaturized Four-element MIMO antenna system for sub-6 GHz Fifth-generation (5G) RF device applications generating different bands and high port isolation properties is presented in this work.
- The isolation of greater than 18 dB is seen between any two radiating elements.
- The proposed antenna features a new configuration, higher isolation, lower CCL values, and higher efficiency values as compared to the ones stated, and the structure is developed exclusively for 5G sub-6 GHz applications.

7.1.4 Inferences from a novel shaped hexa-band frequency reconfigurable antenna

- The designed antenna has many advantages due to its characteristics such as light weight, compact size, low-cost, easy to fabricate and also simple in integration.
- The outstanding efficiencies, satisfactory radiation patterns, acceptable gain, better impedance matching and directivity values were achieved for the different operating conditions of the proposed hexa-band antenna.
- The recommended antenna can be intended for use in the satellite communications, military purposes and in modern communication devices such as laptops and tablets.

7.2 Future Perspective

The thesis provides a new concept of designing MIMO antennas for different applications in the future with a single ground. In the MIMO antennas, the direct split in the ground plane can yield a direct enhancement in port isolation and hence there is a very low current coupling through the ground plane occurs. In addition, DGS technique can be used to enhance the port isolation in MIMO antennas and multi-antenna systems is still valid if the ground plane is not totally split. The proposed antenna has also shown good agreement between both the simulated and measured results. These antennas could be fabricated by using other substrates to explore their scope in the area of wearable antennas or in medical applications. The assessment of output results has shown good agreement in parallel to the simulated results, thus solidifying the validity of the proposed prototypes. Further analysis of Hexa-band frequency reconfigurable antenna can be used with another switching techniques in future. It can also be done by using another substrate for satellite or military applications.

www.ingramcontent.com/pod-product-compliance
Lightning Source LLC
LaVergne TN
LVHW082334200926
848481LV00012B/3342
* 9 7 9 8 2 3 2 0 3 1 5 7 2 *